Tourism and Hospitality for Sustainable Development

Emmanuel Ndhlovu • Kaitano Dube
Ngoni Courage Shereni
Editors

Tourism and Hospitality for Sustainable Development

Volume One: Technological Innovations and Development Realities

Springer

Editors
Emmanuel Ndhlovu
Vaal University of Technology
Vanderbijlpark, South Africa

Kaitano Dube
Vaal University of Technology
Vanderbijlpark, South Africa

Ngoni Courage Shereni
School of Tourism and Hospitality
University of Johannesburg
Johannesburg, South Africa

ISBN 978-3-031-63068-2 ISBN 978-3-031-63069-9 (eBook)
https://doi.org/10.1007/978-3-031-63069-9

This Springer imprint is published by the registered company Springer Nature Switzerland AG
The registered company address is: Gewerbestrasse 11, 6330 Cham, Switzerland

About the Book

This book examines digitalisation and sustainable development in the tourism and hospitality sectors from theoretical, empirical, and policy perspectives. It shows that digitalisation technologies increasingly influence the products and services that tourism and hospitality enterprises prefer and offer globally. The prospects of new technologies to completely overhaul the tourism and hospitality industry can no longer be denied. New generations of tourists have different expectations and requirements in relation to today's tourists. On the other hand, enterprises now offer tourism and hospitality products and services differently. The sharing economy and the experience economy strongly influence the creation of new business models in tourism. Tourism and hospitality enterprises can potentially benefit from this digital transformation. Enterprises that resist change risk being left behind.

The book comprises contributions that generate theoretical insights, empirical findings, and evidence-based recommendations by focusing on emerging and forecasted technologies that are used in the tourism and hospitality industry, such as blockchain, robots, artificial intelligence, virtual reality, and big data analytics, among others. This technology diffusion is studied from various angles, including topics that include Blockchain Technology: Drivers for and Barriers to Adoption (Organisational issues related to the adoption of blockchain technology in the business world; financial issues; strategic and technical issues); Big Data, Analytics and Ethical Considerations (Ethical issues and social responsibility as they relate to technology adoption; and the risks generated by rapid or radical technological adoption in tourism and hospitality enterprises regarding ethical data management).

Contents

About the Contributors

O. A. Ajigini is presently with The Independent Institute of Education at the Faculty of ICT. He has written many journal papers and delivered papers at international conferences. He holds a PhD from the University of South Africa, a Master of Information Technology degree from the University of Pretoria, and an Honours BSc in Information Systems. His research areas include digital transformation, Technology Acceptance modelling, the Internet of Things, and cybersecurity.

Ilhaam Banoobhai-Anwar is a Lecturer, Founder, and Entrepreneur with over 15 years of experience at the Cape Town Hotel School in the Faculty of Business and Management Sciences at the Cape Peninsula University of Technology in Cape Town, South Africa. Her research and teaching interests include e-commerce, hospitality and tourism, gastronomy, and nutrition.

Tendai Chibaya holds a PhD in Economics and Management Sciences with Tourism Management from North-West University, South Africa. She lectures in the Department of Hospitality, Tourism and Culture at Great Zimbabwe University, Zimbabwe. Her research interests are event tourism management, cultural and heritage tourism, disability tourism, and tourism management.

Peter Chihwai (PhD) is a Postdoctoral Fellow at the Vaal University of Technology with experience as a researcher, lecturer, marketer, manager, and public relations practitioner. He has published in peer-reviewed journals. He has published book chapters and a book titled *Tourism Marketing and Management*. His research areas include tourism marketing, tourism marketing, strategy, management, and service quality.

Desré Draper is a Senior Lecturer at the Cape Town Hotel School, Cape Peninsula University of Technology, South Africa. Her research and teaching focus areas include Rooms Division, Revenue Management and Hospitality Information Systems.

Kaitano Dube is an Associate Professor of Tourism Geography seconded to the Ecotourism Management Department. He hosts Postdoctoral Fellows and is acting as a Faculty of Human Science Research, Innovation and Commercialisation Professor. He serves as a senate member in several senate subcommittees such as the Higher Degrees Committee, Senate Research and Innovation Committee Member, and Faculty Research and Innovation Chairman, among others. His role is to promote research, innovation, and commercialisation in the faculty. He is a C NRF rated researcher and a visiting Professor at Emirates Aviation University, EAU.

Tawanda Makuyana is currently attached to the Vaal University of Technology, South Africa, as a Post Doctoral Fellow. He is an accomplished researcher in the following areas: human capital development/human resources issues, vulnerable groups and disability issues (advocacy for socio-economic empowerment of disabled people), inclusive workplace and wellness, inclusive organisational policy and strategy development and implementation, accessible/inclusive destination competitiveness, inclusive community-based tourism, nature (biodiversity)-based tourism, Inclusive sport tourism, and accessible tourism. Mixed-method research design is preferred.

David Mhlanga is a Senior Research Fellow at the University of Johannesburg, South Africa. His research interests include Fintech, financial inclusion, poverty studies, Industry 4.0, and the application of artificial intelligence and machine learning in business economics and finance. His subject areas include development economics, digital economics, economics of artificial intelligence, health, and education economics. He holds a PhD from the North-West University, South Africa.

Tshenolo Montsiemang is an educator and upcoming scholar in the field of tourism. As an educator, he has taught courses at both the undergraduate and graduate levels, inspiring students to pursue careers in tourism and make meaningful contributions. His research interests include tourism policies, tourism development, tourism and sustainability, as well as tourism and culture

Givemore Moyo holds a Bachelor Honours Degree in Banking (National University of Science and Technology), a Postgraduate Diploma in Tertiary Education (Midlands State University), and a Master of Science in Finance and Investment (National University of Science and Technology). The author is a Former Lecturer in the Department of Business Management, Catholic University. He is a Lecturer at Midlands State University in the Department of Accounting Sciences, Harare Campus.

Emmanuel Ndhlovu is a scholar and student of development. His research interests are in peasant and informal sector livelihoods, land reform issues, political economy, digitalisation of industries, and food sovereignty. He holds a PhD in Development Studies from the University of South Africa. At the time of writing, Emmanuel was with the Vaal University of Technology as a postdoc researcher.

Tshinakaho Nyathela-Sunday (PhD) is a Senior Lecturer at Cape Town Hotel School, Faculty of Business and Management Sciences, Cape Peninsula University of Technology. The portfolio, among others, includes a research and internationalisation coordinator. She has worked in various institutions and hospitality sectors across South Africa. She serves in various internal and external institutional committees. She is a South Africa Netherlands Research Programme (SANPAD) awardee and a researcher in hospitality management and related fields. She has participated in various international conferences and reviews for various journals.

Anupama Sethi (PhD) is a Lecturer at the Monash Business School, Monash University, Australia. Her background lies in the area of development economics. She has a PhD, MPhil, and MA in Economics, along with a Bachelor of Education. She is passionate about creating a positive learning environment and making the learning experience intellectually engaging and enjoyable for students. Anupama has been a consistent and significant contributor to economics teaching development and delivery over several years. Anupama's efforts have been recognised with the department teaching awards, including Dean's Citation for Outstanding Contribution to Student Learning and appreciation by her students.

Ngoni Courage Shereni is a Senior Postdoctoral Research Fellow in the School of Tourism and Hospitality at the University of Johannesburg in South Africa. He holds a PhD in Tourism and Hospitality Management from the University of Johannesburg. His research interests are in sustainable tourism, Sustainable Development Goals (SDGs), climate change, tourism education, disruptive technology, urban tourism, tourism exhibitions, community-based tourism (CBT), and community-based natural resource management (CBNRM) practices.

Zibanai Zhou is a Lecturer in the Department of Tourism, Hospitality and Leisure Sciences at Midlands State University, Zimbabwe. His main research interests span tourism development, event tourism, and local communities.

Linnet Zimusi holds a Bachelor's Honours Degree in Finance (Great Zimbabwe University), a Master's in Finance and Investment (National University of Science and Technology), and a Postgraduate Diploma in Tertiary Education (Midlands State University). She is a Lecturer in the Department of Accounting Sciences, Midlands State University, Gweru Campus.

Chapter 1
Introduction: Historicising and Theorising Digital Transformation and Sustainable Development in the Tourism and Hospitality Industry

Emmanuel Ndhlovu, Kaitano Dube, and Ngoni Courage Shereni

1.1 Introduction

It may be partially incorrect to assert that the tourism and hospitality industry, particularly in the twenty-first century, can no longer be imagined, designed, marketed, managed, or studied independently of digital mediation and technology. This is evidenced by the increase in the rate at which digitalisation is adopted by tourism and hospitality establishments in both the Global North and South contexts. Seminal studies argue that digitalisation not only results in improved profitability but also enables businesses to grow sustainably and be positioned to effectively respond to challenges such as climate change, hyperinflation, recurring pandemics, and energy crises, among others (Bisoi et al., 2020; Hassan, 2023; Mhlanga et al., 2021; Youssef & Zeqiri, 2021). Digitalisation has numerous characteristics and forms in the tourism and hospitality industry. Some digital innovations serve as managerial tools, others enable the reorientation of current business models, whereas others have a disruptive effect that revolutionaries the way of doing business (Christou et al., 2020; Filimonau & Naumova, 2020; Youssef & Zeqiri, 2021). This enables them to steer the sustainable development agenda.

Technologies are, however, not a perfect-finished whole or a straightforward process, as the 'human touch' and escapism remain the pull factors for industries, enterprises, or individuals to participate in tourism and hospitality. The opportunities, costs, and challenges differ in geo-political and socio-economic contexts

E. Ndhlovu (✉) · K. Dube
Vaal University of Technology, Vanderbijlpark, South Africa
e-mail: kaitanod@vut.ac.za

N. C. Shereni
University of Johannesburg, Johannesburg, South Africa

E. Ndhlovu et al. (eds.), *Tourism and Hospitality for Sustainable Development*,
https://doi.org/10.1007/978-3-031-63069-9_1

(Ndhlovu & Dube, 2023). Digitalisation and technology integration can be considered two sides of a coin. On the one hand, digitalisation and technology help enterprises deal with competition and thus bolster their sustainability as businesses. Filimonau and Naumova (2020) argue that digitalisation presents novel ways to interact effectively with customers.

Similarly, infusing technology into business operations enables enterprises to reach new customers and improves customer engagement and service provision (Thrasso et al., 2021). On the other hand, technology integration offers new capacities, thus generating competition through digitalisation. Therefore, the industry goes through cycles of sustained transformation, often with some of the changes being disruptive.

Digital innovations and inventions can play a significant role in universal design and reasonable accommodation for all tourism consumers, including improving the accessibility and consumption of tourism products and services by vulnerable and excluded categories such as the disabled, elderly, and poor. This is one of the critical pillars of accessible tourism and hospitality. In response to various challenges businesses face and changing consumer needs, this edited book explores the trends, challenges, opportunities, and complexities of digitalisation, sustainable development, and technology integration in the tourism and hospitality industry, contributing to emerging research streams and practical applications. The aim is to provide insights into how tourism and hospitality businesses embrace technological innovations to support sustainable growth and how digitalisation influences the interaction with various stakeholder groups.

The book is located at the juncture of three sets of literature: literature on tourism and hospitality, literature on sustainable development, and literature on digitalisation. As a common trend, these three strands of literature must adequately engage with each other. Thus, this book addresses this gap through several case studies of various tourism and hospitality enterprises in different parts of the world. Much of the literature on digitalisation in the tourism and hospitality industry is focused on how enterprises integrated digital tools to survive coronavirus (COVID-19) safety protocols. This has tended to crowd out studies about digitalisation in the pre- and post-COVID-19 period. This volume, thus, emphasises the importance of integrating digital technologies in the tourism and hospitality industry before and after COVID-19 in offering a more refined analysis of twenty-first-century digital inventions, innovations, and development realities in the tourism and hospitality industry.

This chapter is organised as follows: The next section reviews the definitions given to the concept of digital transformation. This is followed by a discussion on the history of digitalisation with a focus on the tourism and hospitality industry, drivers of digitalisation, and opportunities and challenges of digitalisation. After that, sustainable development is conceptualised. The research methodology for the book is then summarised. Lastly, the volume outline is presented, followed by an overall conclusion.

1.2 Defining Digital Transformation

A universally accepted definition of digital transformation needs to be created. Much debate and confusion concerning how the term can be precisely defined. As a result, the term has many variants and is often misused in the public discourse. In the current chapter, the term describes and defines the different digitalisation processes and their impact on society and its systems. Caution should be exercised when describing the impacts of digitalisation and technological advancement on society's political, economic, social, and cultural systems. This chapter, therefore, explores the different definitions of digitalisation and proposes a definition that provides a broad conceptualisation of the concept. This is necessary for understanding the global tourism and hospitality industry. This understanding is indispensable as it opens up new categories of knowledge crucial in searching for solutions to the challenges that may emerge from digitalisation. Table 1.1 summarises selected definitions of the concept of digital transformation.

Based on the above conceptualisations, we conceptualise digital transformation as the radical departure from human-controlled activities to mechanical-controlled ones using the novel technologies accompanying the Fourth Industrial Revolution (4IR). As such, there can be no one-size-fits-all definition for the process.

1.3 History of Digital Transformation

Although digital transformation gained prominence immediately before and after the advent of COVID-19, digital tools were already applied across different industries from the 1990s and even much earlier. For instance, in the mass media and retail sectors, digital channels were considered essential strategies to reach customers quickly and effectively in the 1990s and 2000s (Schallmo et al., 2017). This was even though purchases were still conventionally made using cash and within the stores. Online purchases using real-time innovative tools and digital money transfers did not exist by then. The digital revolution builds on previous advances in research and innovation stretching back to the 1940s when Claude Shannon published the article: *A Mathematical Theory of Communication* (Shannon, 1948). Today, Shannon is arguably accepted as the father of modern digital technology.

The years from 2000 to 2015 saw a rapid rise in smart devices and social media use by individuals and businesses. This drastically transformed not only business operations but also customer perceptions and preferences. Digital tools defied time and space and transformed how customers interacted with businesses. Communication between the business world and its clients has shifted to digital platforms, making it possible to offer real-time customised interactions. A significant development in the late 2000s was the growth of digital payment platforms such as PayPal, which popularised online commerce and increased opportunities for web-based points of sale. These payment methods have become popular across

Table 1.1 Existing definitions of digital transformation

Author(s)	Definition
Schallmo et al. (2017: 3)	"The DT framework includes the networking of actors such as businesses and customers across all value-added chain segments and the application of new technologies. As such, DT requires skills that involve extracting and exchanging data and analysing and converting that data into actionable information. This information should be used to calculate and evaluate options to enable decisions and/or initiate activities. In order to increase a company's performance and reach, DT involves companies, business models, processes, relationships, products, etc".
Bowersox et al. (2005: 22)	Digital Business Transformation (DBT) is a "process of reinventing a business to digitise operations and formulate extended supply chain relationships. The DBT leadership challenge is about re-energising businesses that may already be successful in capturing the full potential of information technology across the total supply chain."
Westerman et al. (2011: 5)	"DT—the use of technology to radically improve the performance or reach of enterprises—is becoming a hot topic for companies across the globe. Executives in all industries are using digital advances such as analytics, mobility, social media, and smart embedded devices—and improving their use of traditional technologies such as ERP—to change customer relationships, internal processes, and value propositions."
Mazzone (2014: 8)	"DT is the deliberate and ongoing digital evolution of a company, business model, idea process, or methodology, both strategically and tactically."
BMWi (2015: 3)	"Digitisation stands for the complete networking of all sectors of the economy and society and the ability to collect relevant information and analyse and translate that information into actions. The changes bring advantages and opportunities, but they create completely new challenges".
PwC (2013: 9)	"DT describes the fundamental transformation of the entire business world by establishing new technologies based on the internet that have a fundamental impact on society".
Boueé and Schaible (2015: 6)	"We understand DT as a consistent network of all sectors of the economy and the adjustment of the players to the new realities of the digital economy. Decisions in networked systems include data exchange and analysis, calculation and evaluation of options, initiation of actions and the introduction of consequences".

Source: Authors; Developed from various sources

industries, including tourism and hospitality, where mobile devices and other disruptive technologies capture customers' online activities (Shereni & Chambwe, 2020). Tourism and hospitality enterprises use this personalised information better to tailor their products, communications, and interactions to fit customers' needs. Figure 1.1 summarises the history of digital transformation.

Digitalisation in the tourism and hospitality industry shifted from traditional business models to e-business, e-commerce, and smart tourism, as shown in Fig. 1.2.

The last stage in Fig. 1.2 could have commenced around the 2000s and is still unfolding. During this period, a steady rise of digitalisation began to gain inroads in industries, including tourism. The term 4IR is commonly used to describe the digital explosion we are currently witnessing, characterised by the widespread development of various digital innovations in the domain of big data analytics, artificial

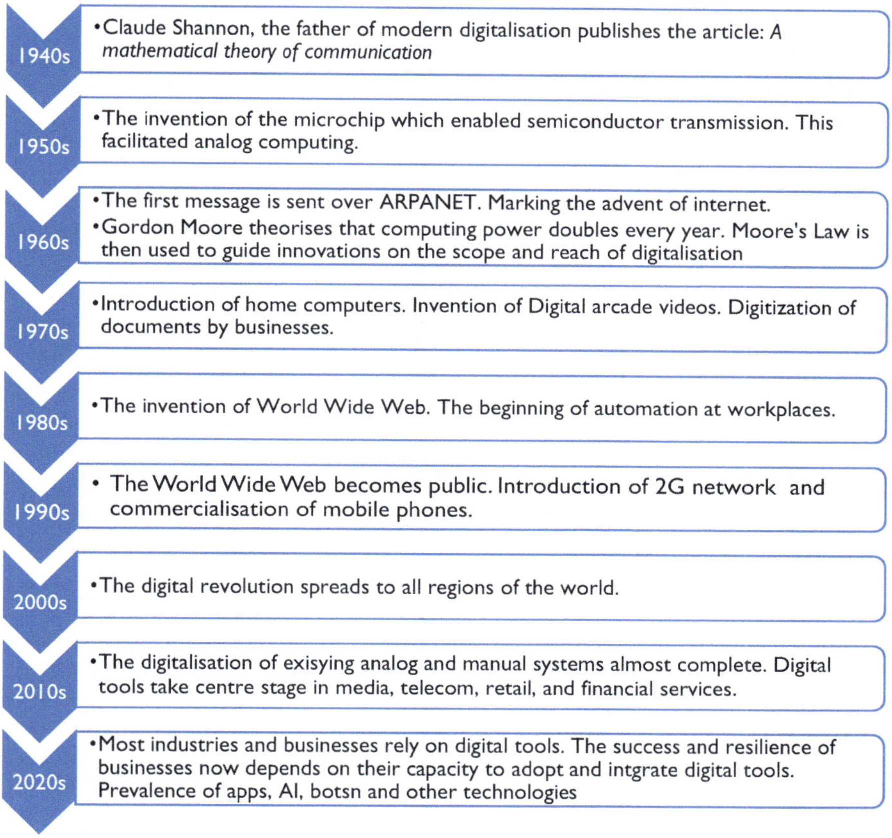

Fig. 1.1 History of digital transformation. (Source: Authors, Developed from various sources)

Fig. 1.2 Digital transformation in tourism. (Source: Authors, Developed from several sources)

intelligence, virtual reality, and cloud computing, among others (Marx et al., 2021; Youssef & Zeqiri, 2021). The 4IR and its related variants, like Business 4.0, have helped transform business practices and are critical in reshaping economic production in various industries (Xiang & Fesenmaier, 2017). Tourism as a service industry is experiencing enormous transformation due to technological advancements and the digitalisation processes we are currently witnessing.

Over time, digitalisation has led to three significant phases of technological advancements in the tourism industry. The initial phase occurred between 1990 and 2000 and was primarily focused on sales and marketing. This phase was closely associated with the rise of the Internet, which enabled businesses to incorporate technology into their marketing strategies (Gorenšek & Kohont, 2019). Moreover, digital software emerged, providing small enterprises with tools to enhance their internal operations (Kindzule-Millere & Zeverte-Rivza, 2022). During this phase, websites gradually replaced traditional paper-based marketing materials, web-based reservation systems facilitated business transactions, and distribution structures greatly improved coordination within the industry. Notable technological advancements during this period included electronic cash registers, financial software, mobile phones, emails, the intranet, internet banking, office software, video conferencing, and websites.

The second phase, between 2000 and 2010, witnessed the ascent of digital business ecosystems. Within this timeframe, tourists progressively relied on the Internet as a primary source of information. Technological advancements catalysed the establishment of virtual marketplaces, enabling online transactions to transpire (Dredge et al., 2018). Furthermore, the focal point shifted from product orientation to prioritising valuable customer experiences and cultivating personal client relationships (Fernindez-Rovira et al., 2021). This era also witnessed the emergence of online intermediaries like Expedia and review platforms like TripAdvisor, granting customers greater autonomy in their purchasing decisions (Dredge et al., 2018). Consequently, visitor information centres diminished significantly as tourists sought assistance from online platforms. Notable tools and products during this phase encompassed smartphones, computer graphics software, property management systems, computerised ticketing systems, stock control systems, online booking systems, customer reservation systems, email marketing, and customer relationship systems.

The third phase, the Integration of Systems, started around 2010 and is still ongoing. During this phase, there have been advancements in various areas, including cloud computing, mobile applications, the Global Point System, virtual reality, augmented reality, Google Analytics, and wearable technologies (Dredge et al., 2018; Mhlanga et al., 2021). The rise of social media platforms and Web 2.0 has allowed for the emergence of user-generated content, opening up opportunities for innovative product development. Additionally, collecting customer data from different platforms has enabled tourism enterprises to effectively customise their offerings by analysing customers' digital footprint.

1.4 Drivers of Digitalisation

Various factors influence enterprises' adoption and integration of digital technologies (Kindzule-Millere & Zeverte-Rivza, 2022; Mazzone, 2014). These drivers of digitalisation encompass marketplace influences, competition, social and demographic characteristics, economic conditions, geophysical environments, political settings, and awareness of diverse technologies and legal structures (Hassan, 2023; Marx et al., 2021). Some scholars argue that policy initiatives should consider these drivers rather than simply replicating practices from other destinations (Dredge et al., 2018). One crucial driver of digitalisation in tourism enterprises is the market. In a highly globalised world, markets have become indispensable for the survival and profitability of an enterprise. Therefore, technological advancements in the Fourth Industrial Revolution era rapidly drive changes in customer demands and preferences. Enterprises adopt and integrate digital tools to enhance their market presence. The exponential growth of social media and mobile applications has highlighted the importance of an online presence for reaching and connecting with digitally savvy clients (Hansen & Sia, 2015). Technologically savvy clients monitor their favourite brands online and expect to find their product offerings on digital platforms before visiting a physical store.

Digitalisation is also pursued for economic reasons. It provides opportunities for the creation of novel value-creation processes and the expansion of existing ones. However, the global and local economic conditions affect the speed and extent of digital transformation. Developed countries' enterprises are rapidly digitalising compared to enterprises in developing countries (Ndhlovu & Dube, 2023).

Dredge et al. (2018) have observed that social and demographic factors are crucial in driving digital transformation. Research has consistently shown that children and young people are early adopters of new technologies (Kusuma et al., 2020). Multiple studies have established a strong association between age and technology adoption (Eze et al., 2018; Grewal et al., 2018; Nguyen et al., 2020). Meggiolaro (2018) argues that age is closely linked to employees' performance in IT-based tasks. Younger business owners and employees are more inclined to take risks (Li & Tang, 2010). Anže et al. (2021) Note that executives under 39 are more willing to embrace mobile marketing technologies. More than 70% of these executives utilise mobile devices as communication tools, compared to only 28% of executives aged 50–60. These findings suggest that younger executive managers are more likely to adopt digital technologies compared to their older counterparts swiftly. Likewise, Enze et al. (2021) observed that small enterprises employing young individuals tend to be more innovative. It can be argued that younger employees are more inclined to engage with digital technologies and are more open to new experiences (Jabło´nski & Jabło´nski, 2019). While older business owners and employees do possess physical, social, and mental strengths and can be innovative and dedicated (Troshani et al., 2018), younger business owners and employees generally display a greater willingness to embrace innovation and change (Ransbotham et al., 2019). Consequently, technologies are more likely to be adopted and integrated into enterprises where the owners and employees are enthusiastic about utilising new tools.

Digitalisation is also driven by the nature of politics and legal frameworks that play out in particular settings. The political environment often directs the policy framework for developing the right conditions for the uptake of technologies. Likewise, legal frameworks are crucial in safeguarding business and interests and guiding data security and privacy issues.

Competitors also influence the need for innovation and the uptake of digitalisation. In a fast-globalising world, business organisations must keep pace with technological changes to maintain competitiveness (Nguyen et al., 2020). Okundaye et al. (2019) note that enterprises integrate novel technologies to respond to competition. Alabadi et al. (2019) aver that enterprises that adopt technology gain a competitive advantage. Opute et al. (2020) argue that businesses adopt technologies to stay in the market and have a competitive advantage by offering their products online before their competitors, creating a competitive edge over the competitors in the market. Dredge et al. (2018) note that disruption from new competitors drives the conceptualisation of new and innovative business models, necessitating the adoption of novel technologies.

Another factor that drives digitalisation by enterprises is the availability, accessibility, and awareness of different technologies, as well as opportunities for training on new technological developments (Kusuma et al., 2020). Enterprises will likely adopt and integrate digital technologies when the infrastructure is readily available and affordable (Ndhlovu & Dube, 2023). Kamaruzaman et al. (2019) note that infrastructural insufficiency is another main challenge that frustrates technological transition in developing countries. This is supported by Kansakar et al. (2019), who emphasised that the successful adoption of digital technologies requires the availability of relevant technological infrastructures.

Geographical location and physical features also powerfully drive the adoption of technology. The geographic location significantly affects the accessibility of tourism establishments to sufficient digital infrastructures. Modern wireless and fibre broadband connections are readily available for businesses in urban areas, unlike rural and remote areas where such facilities may be inaccessible (Dredge et al., 2018). Rocha et al. (2016) found that in most countries in the Global South, the concentration of digital infrastructure is mostly in urban areas and very little in rural areas. This was found to have increased the 'digital gap' in countries such as Malawi, Senegal, Kenya, India, and Brazil while depriving tourism businesses located in remote areas of opportunities to adopt digital technologies (Cirera et al., 2022). That being said, the factors that drive digitalisation by businesses can be either benefits or barriers or both, as discussed in the next section.

1.5 The Opportunities and Challenges of Digitalisation

Since the latter part of the last decade, there have been rapid, disruptive, and radical changes in business operations, culture, and knowledge production. With digital tools and platforms (such as Facebook, Twitter, and Wikipedia) now assuming centre stage, the world is seeing the emergence of new non-proprietary and non-marketable tools that aid in producing knowledge. This is precisely due to the exclusive accessibility of digital tools, which have changed how people interact (Hassan, 2023). Since the digital revolution has brought about enormous opportunities for business growth and smooth operations, it has also spewed considerable challenges due to its disruptive nature. These challenges and dangers, therefore, need to be highlighted in the analysis.

The reviewed literature agrees that digitalisation is transforming the organisational environment to the extent that the effects are manifested across industries and sectors in terms of expectations and desires (Bisoi et al., 2020; Gorenšek & Kohont, 2019; Ndhlovu & Dube, 2023). Digitalisation also steered various changes that have enabled organisations to grow profitably. Figure 1.3 summarises the opportunities brought about by digital transformation.

In the context of the tourism and hospitality industry, digitalisation stimulates new business models, leads to creativity and innovation, improves visitor satisfaction, enables increased customisation of visitor experiences, contributes to new destination configurations, new business ecosystems, new value chains, and introduces up new roles for stakeholders such as consumers and producers (Filimonau & Naumova, 2020; Hassan, 2023; Kindzule-Millere & Zeverte-Rivza, 2022).

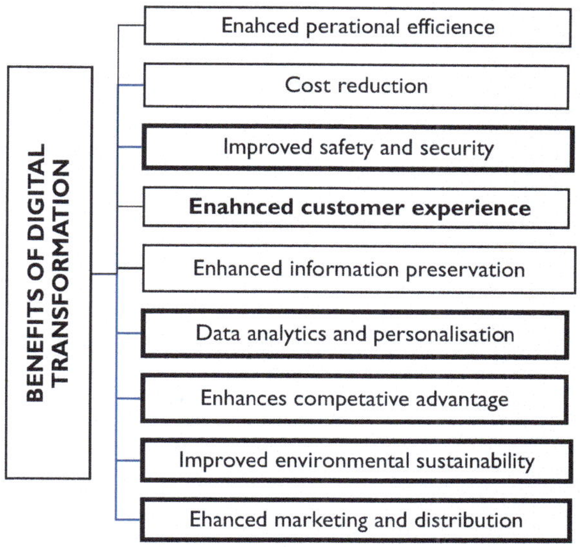

Fig. 1.3 Benefits of digital transformation. (Source: Authors, Developed from various sources)

Across industries in general, digitalisation has resulted in the coming together of different industry sectors. This has seen certain professions becoming redundant while at the same time creating new ones (Gökkaya, 2020; Kindzule-Millere & Zeverte-Rivza, 2022). However, this transition brings about its challenges. For example, there will be an urgent need for training and retraining employees to be equipped with the skill set required for those professions (Bozhuk et al., 2020; Fernindez-Rovira et al., 2021).

The digitalisation of documents represents the first digitisation stage in operational processes. As recently as 2019, digital transformation was estimated to contribute more than US$1 trillion to the Asia-Pacific region's Gross Domestic Product (GDP) by 2021 (Gorenšek & Kohont, 2019). Although no study has followed up on this estimation to confirm the projection, this highlighted the central role of AI as the primary facilitator for business growth. However, among the many challenges organisations have reported regarding digitalisation are digital competencies and cyber-security (Cuomo et al., 2021; Fernindez-Rovira et al., 2021). The tourism and hospitality industry is very fragmented, and its sub-sectors, such as restaurants, accommodation, catering, transport, and personal services, face opportunities and challenges regarding digital transformation (Gokkaya, 2020). There are substantial variances in human resource aptitudes, different access levels to financial and non-financial resources, awareness, and differences in digital skills (Mhlanga et al., 2021).

The skills and competencies needed to adopt digitalisation are sometimes complex; for example, advanced analytic tools may not be available to enable organisations to respond efficiently to market demands (Marx et al., 2021). Digitalisation provides exceptional knowledge access and reduced operational costs (Shimmura et al., 2020; Solnet et al., 2019). Another challenge scholars raise is the problem of the proliferation of forms of work in the industry. This proliferation is characterised by the blend of ordinary forms of work and precarious work with illegal status and informal forms of employment in the sector (Gorenšek & Kohont, 2019). These can include informal tourist guides, vendors, and photographers (Ndhlovu & Mhlanga, 2023). These problems are linked to the trend of digitalisation, which has made the instant processing of photographs and money and airtime transfer possible. While this directly affects formal businesses and their employees, it also affects customers increasingly exposed to illegal stakeholders. Digitalisation also changes interpersonal relationships between employees within enterprises and customers (Berger & Frey, 2016).

1.6 Theorising Sustainable Development

Sustainable tourism is economically viable in terms of keeping the resources on which the future of tourism depends (the physical environment and the social fabric of host communities) (Swarbrooke, 2000). The critical elements around sustainable tourism include the maintainable use of resources, decrease in resource consumption, waste reduction, preserving diversity, supporting local economies, involvement of local economies, stakeholder participation, and responsible marketing (Mudasir et al., 2020). These fundamental elements emphasise the need to balance socio-cultural and environmental aspects.

The concept of 'sustainable tourism' directly emerges from the idea of sustainable development steeped in the global discussions encapsulated in the Bruntdland Report. Sustainable development is anchored on three pillars: economic prosperity, environmental conservation, and social equity (Zhang & Shang, 2022). The Bruntdland Report defines sustainable development as "development that meets the needs of the present generation without comprising the ability for future generations to meet their own needs" (World Commission on Environment and Development, 1987: p.34). The definition identifies fundamental principles of sustainability such as environmental protection (biodiversity) and artificial heritage; preservation of vital ecological processes; a holistic approach to planning and strategy; facilitation of public participation; production to be pursued in a manner that does not endanger the needs of future generations; and equality and opportunity between different countries (Guan et al., 2022).

In the context of tourism, sustainable tourism follows the application of the principles of sustainable development to the tourism industry (Shereni, 2022). Thus, implementing sustainable tourism principles requires optimum use of environmental resources, vital sustenance, and conservation of natural and cultural heritage. It also requires respecting the socio-cultural legitimacy of host communities and preserving their built and living cultural heritage and traditional values (Koçak et al., 2020). In addition, implementing sustainable tourism requires tourism businesses to contribute to intercultural understanding and tolerance and ensure sustainable, long-term economic activities, fairly sharing socio-economic benefits with all stakeholders. This includes jobs, other income-earning opportunities, community social services, and poverty reduction (Abdou et al., 2020).

Sustainable tourism is characterised by socially responsible tourism, the prospect for all citizens to participate in tourism, job opportunities, and maximising benefits to the host communities (Abouzeid, 2022). Some scholars have found the principles of sustainable development to be relatively rarely applied in tourism enterprises globally (Zhang & Shang, 2022). According to Mudasir et al. (2020), tourism enterprises often ignore the social and environmental costs of doing business in pursuit of economic profit. Some of the barriers to the adoption of sustainability ethos include a lack of awareness (Ndhlovu & Dube, 2023), a general lack of conviction (Mudasir et al., 2020), lack of education and training by staff (Ndhlovu & Dube, 2023) as well as misplaced budgetary priorities (Shereni et al., 2023).

The chapters in this book show that in recent decades, efforts towards the realisation of sustainable tourism are increasingly being linked to Fourth Industrial Revolution (4IR) technologies, such as the Internet of Things (IoT), artificial intelligence (AI), cyber-physical systems (CPS), virtual reality (VR), and augmented reality (AR) which scholars consider as having much potential to reduce or monitor CO_2 emissions which cause climate change (Friedrich et al., 2020; Gossling et al., 2023; Narayan et al., 2022). However, although these technologies have widely been embraced at policy and academic levels, gaps still need to be found in how stakeholders within tourism and hospitality enterprises perceive their deployment.

1.7 Methods and Materials

The chapters in this volume utilise various research methods to comprehensively understand digitalisation in the tourism and hospitality industry and its impact on businesses, employees, customers, and the environment. Contributors employ quantitative, qualitative, and mixed methods research methodologies to yield detailed discussions on digitalisation. Qualitative methods include in-depth interviews, document analyses, focus groups, systematic literature reviews, and descriptive analyses. Quantitative methods include the application of the PRISMA approach, bibliometric analysis, and sentiment analysis. Implementing these diverse methodologies ensures meaningful contributions to both practical and theoretical domains. This book, the first in three volumes, provides theoretical insights, empirical findings, and research-based recommendations. The book focuses on emerging and projected technologies in the tourism and hospitality industry, such as blockchain, robots, artificial intelligence, virtual reality, and big data analytics. Technology diffusion is examined from various perspectives, including blockchain technology, big data analytics, financial technology, the sharing economy, inclusive tourism, customer experiences, and barriers to adopting digital technologies.

1.8 Volume Outline

The book has 13 chapters, including this introductory chapter, which theorises and historicises digitalisation and sustainable development in the global tourism and hospitality industry. The following 13 empirical chapters appear in four parts: *Introduction and background*; *Conceptualisation and review of digital technologies in tourism; Barriers to and challenges technology adoption in tourism businesses; and Policy and practical recommendations.*

Part I comprises one chapter conceptualising digitalisation in the tourism and hospitality sectors. Chapter 2 by Tawanda Makuyana, Emmanuel Ndhlovu, and Kaitano Dube conceptualises and explores digital transformation in the tourism and hospitality sectors' training and training. The chapter shows that technical universities, polytechnics, hotel schools, and liberal universities that offer tourism and hospitality programmes adopt various digital-based teaching and learning as driven by a variety of contexts, like educators' digital competencies, accessibility of digital learning platforms among learners based on socio-economic backgrounds and location, resources base to support the learning needs among various learners, diversity in students' learning process, co-creation of learning environment, assessment, engagement, and motivation of learners.

Part II comprises four chapters providing detailed case study-based accounts of digitalisation within tourism and hospitality enterprises. Chapter 3 by Peter Chihwai explores the digital innovations adopted by South Africa National Parks (SANParks), Southern Sun hotels, air transport, and the airports industry. In Chap. 4, Givemoyo Moyo, Linnet Zimusi, and Emmanuel Ndhlovu report that WhatsApp, Facebook, computerised billing and receipting, POS, and mobile money services are some of the digital tools used in female-owned restaurants in the Harare Central Business District, Zimbabwe. This is followed by Chap. 5 by Ngoni Courage Shereni and Emmanuel Ndhlovu, which explores the prospects of utilising peer-to-peer platforms in urban tourism in Zimbabwe. The investigation needs to show more use of these platforms in Zimbabwe. Lastly, Chap. 6 by Emmanuel Ndhlovu, Tawanda Makuyana, and Kaitano Dube explores the status of digitalisation within global aviation enterprises. The study found a differentiated degree of digitalisation and technology integration across airlines.

Part III comprises three chapters that describe the barriers to and challenges of digitalisation by tourism and hospitality enterprises. In Chap. 7, Zibanai Zhou and Tendai Chibaya investigated the opportunities and challenges of harnessing digital technologies in wildlife tourism resource conservation in Zimbabwe. The study identified drones, mobile gadgets, surveillance cameras, collar tags, and remote sensing as digital tools harnessed in wildlife resource conservation across Zimbabwean national parks. The challenges identified in deploying digital technologies encompass inadequate financial resources, lack of technical expertise, and vandalism of infrastructure. Chapter 8 by Tawanda Makuyana, Emmanuel Ndhlovu, and Kaitano Dube examined the impacts of digitalisation on value co-creation in restaurant service encounters among disabled people in Sub-Saharan Africa. They found that digitalisation has mixed value co-creation in restaurant service encounters among various disabled people in the region due to the diversity of impairments, their digital support/needs during service encounters, and their attitudes and level of digital inclusion. The study concludes that digitalisation does not uphold disability inclusion unless it embeds a universal design with the context of the type of restaurants and their access-need target market segments. Tshinakaho Nyathela-Sunday, Ilhaam Banoobhai-Anwar, and Desré Draper, in Chap. 9, explore the challenges of hotel accommodation booking technologies from a management perspective. The study shows that although hotel managers strive to increase direct bookings, most reservations come from external third-party sources. This can be due to not knowing the costs of acquiring direct bookings, hotel websites being managed externally, and OTA's offerings and benefits. Several aspects challenge hotel management in receiving direct bookings, and this needs to be addressed to ensure hotels' competitiveness with other counterparts.

Part IV comprises three chapters that provide directions on the practicality of digitalisation in the tourism and hospitality sectors. The chapters also sketch some policy recommendations. In Chap. 10, Emmanuel Ndhlovu, Kaitano Dube, and Ngoni Courage Shereni explore the strategies that may be used to promote the adoption and integration of technology by small hospitality and tourism enterprises. The chapter concludes that national governments should be the key stakeholders in promoting enterprise technology adoption and integration. In Chap. 11, Emmanuel Ndhlovu and Tshenolo Montsiemang explore the challenges of digital digitalisation and technological integration in Africa. The chapter shows that technology integration results in security risks about data and fraud, increased risk of job cuts, overdependence on technology, and increased costs for small enterprises. The chapter concludes that although technology wields enormous benefits for businesses, the impact can adversely affect enterprises and employees in emerging economies such as Africa. Chapter 12, by David Mhlanga and Anupama Sethi, explores the utilisation of Fintech in Asia to revitalise the tourism industry in the aftermath of the pandemic.

Additionally, the chapter explores the potential repercussions and opportunities that may arise for Africa. They conclude that adopting financial technology (Fintech) is crucial for Africa to regain its dominance in the tourist industry before the pandemic and establish a progressive and forward-looking trajectory in this sector. Lastly, in Chap. 13, Emmanuel Ndhlovu, Kaitano Dube, and Ngoni Courage Shereni summarise the book, draw conclusions from the discussions by contributors, and make practical and policy recommendations.

1.9 Conclusion

This chapter introduces the book Tourism and Hospitality for Sustainable Development—Volume One: Technological Innovations and Development Realities. Its purpose is to provide a historical and theoretical overview of the digital transformation and sustainable development in the tourism and hospitality industry. The chapter emphasises the current global technological revolution and highlights the tourism and hospitality sectors as pioneers in adopting and integrating digital technologies. The ongoing digital revolution, driven by the challenges of the COVID-19 pandemic, has significantly reshaped business practices and interpersonal interactions. These digital trends have profoundly impacted the entire tourism value chain. The chapters in this book offer detailed insights into these technological trends and address the challenges and opportunities associated with adopting and integrating digital technologies in tourism and hospitality enterprises.

References

Abdou, A. H., Hassan, T. H., & El Dief, M. M. (2020). A description of green hotel practices and their role in achieving sustainable development. *Sustainability, 12*, 9624. https://doi.org/10.3390/su12229624

Abouzeid, N. (2022). The impact of digitalization on tourism sustainability: Comparative study between selected developed and developing countries. In *2nd international conference on advanced research in management, economics, and accounting, 18–20 February 2022*. Milan, Italy.

Alabadi, H. F., Sultan, Y. H., & Alkaseer, N. A. (2019). Integration of strategic wellness and sustainable marketing to achieve competitive advantage in the Tourism Industry: An empirical study in a sample of Iraqi tourism agencies. *African Journal of Hospitality, Tourism and Leisure, 8*(5), 1–12.

Anže, Z., Install, J., Jereb, M., & Repo, T. (2021). Sustainable and selective modern methods of noble metal recycling. Angewandte Chemie (International Edition), 62, e202214453.

Berger, T., & Frey, C. B. (2016). Structural transformation in the OECD: Digitalisation, deindustrialisation and the future of work. *OECD Social, Employment and Migration Working Papers, 193*, 1–52.

Bisoi, S., Roy, M., & Samal, A. (2020). Impact of artificial intelligence in the hospitality industry. *International Journal of Advanced Science and Technology, 29*(5), 4265–4276.

BMWi. (2015). *Industrie 4.0 und Digitale Wirtschaft — Impulse für Wachstum, Beschäftigung und Innovation*. Bundesministerium für Wirtschaft und Energie.

Boueé, C., & Schaible, S. (2015). *Die Digitale Transformation der Industrie*. Studie: Roland Berger und BDI.

Bowersox, D. J., Closs, D. J., & Drayer, R. W. (2005). The digital transformation: Technology and beyond. *Supply Chain Management Review, 9*(1), 22–29.

Bozhuk, S., Pletneva, N., Maslova, T., & Evdokimov, K. (2020). *Problems of transformation in the tourism industry in the digital economy*. SHS Web of Conferences 73, 0100, IES201, pp. 1–9.

Christou, P., Similidou, A., & Stylianou, M. C. (2020). Tourists' perceptions regarding the use of anthropomorphic robots in tourism and hospitality. *International Journal of Contemporary Hospitality Management, 32*(11), 3665–3683.

Cirera, X., Comin, D., & Cruz, M. (2022). *Bridging the technological divide: Technology adoption by firms in developing countries*. World Bank Group.

Cuomoa, M. T., Tortorab, D., Foroudic, P., Giordanod, A., Festae, G., & Metallof, G. (2021). Digital transformation and tourist experience co-design: Big social data for planning cultural tourism. *Technological Forecasting and Social Change, 162*(120345), 1–9.

Dredge, D., Phi, G., Mahadevan, R., Meehan, E., & Popescu, E. S. (2018). *Digitalisation in tourism: In-depth analysis of challenges and opportunities. Low-value procedure GRO-SME-17-C-091-A for Executive Agency for Small and Medium-sized Enterprises (EASME) virtual tourism observatory*. Aalborg University.

Eze, S. C., Olatunji, S., Chinedu-Eze, V. C., & Bello, A. O. (2018). Key success factors influencing SME managers' information behaviour on emerging ICT (EICT) adoption decision-making in UK SMEs. *The Bottom Line, 31*(3\4), 250–275.

Eze, S. C., Hart, O., Awa, V. C. A., Chinedu-Eze, & Bello, A. O. (2021). *Humanities and Social Sciences Communications, 8*(82), 1–11. https://doi.org/10.1057/s41599-021-00762-5

Fernindez-Rovira, C., Valdes, J. I., Mollevk, G., & Nicolas-Sans, R. (2021). The digital transformation of business. Towards the datafication of the relationship with customers. *Technological Forecasting and Social Change, 162*, 1–12.

Filimonau, V., & Naumova, E. (2020). The blockchain technology and the scope of its application in hospitality operations. *International Journal of Hospitality Management, 87*, 1–8.

Friedrich, J., Stahl, J., Hoogendoorn, G., & Fitchett, J. M. (2020). Exploring climate change threats to beach tourism destinations: Application of the Hazard–activity pairs methodology to South Africa. *American Meteorological Society, 12*, 529–544. https://doi.org/10.1175/WCAS-D-19-0133.1

Gokkaya, S. (2020). Digital transformation and marketing in tourism industry. In *Selected academic studies from Turkish Tourism Sector, chapter: 11* (pp. 149–164). Peter Lang GmbH..

Gorenšek, T., & Kohont, A. (2019). Conceptualization of digitalization: Opportunities and challenges for organizations in the Euro-Mediterranean area. *Tilen Gorenšek, Andrej Kohont, 11*(2), 93–115.

Gossling, S., Balas, M., Mayer, M., & Ya-Yen, S. (2023). A review of tourism and climate change mitigation: The scales, scopes, stakeholders and strategies of carbon management. *Tourism Management, 95*(2023), 104681. ling).

Grewal, D., Ahlbom, C. P., Beitelspacher, L., Noble, S. M., & Nordfält, J. (2018). In-store Mobile phone use and customer shopping behaviour: Evidence from the field. *J Market, 82*(4), 102–126.

Guan, H., Liu, H., & Saadé, R. G. (2022). Analysis of carbon emission reduction in International Civil Aviation through the lens of shared triple bottom line value creation. *Sustainability, 14*, 8513.

Hansen, R., & Sia, S. (2015). Hummel's digital transformation toward Omni-channel retailing: Key lessons learned. *MIS Quarterly, 14*(2), 132–149.

Hassan, S. C. (2023). Innovative marketing strategies during tourism recovery in the post-COVID-19 period in Nigeria. In P. M. A. S. A. H. J. Kennell (Ed.), *Crisis management, destination recovery and sustainability* (pp. 67–76). s.l.:Routledge.

Jabłonski, M., & Jabłonski, A. (2019). Social factors as a basic driver of the digitalization of the business models of railway companies. *Sustainability, 11*, 3367. https://doi.org/10.3390/su11123367

Kamaruzaman, F. M., Hamid, R., Mutalib, A. A., & Rasul, M. S. (2019). Conceptual framework for the development of 4IR skills for engineering graduates. *Global Journal of Engineering Education, 21*(1), 54–61.

Kansakar, P., Munir, A., & Shabani, A. (2019). Technology in hospitality industry: Prospects and challenges. *IEEE Consumer Electronic Magazine, 8*(3), 60–65.

Kindzule-Millere, I., & Zeverte-Rivza, S. (2022). *Digital transformation in tourism: opportunities and challenges*. Proceedings of the 2022 international conference "Economic Science for Rural Development" No 56 Jelgava, LLU ESAF, 11–13 May 2022, pp. 476–486 https://doi.org/10.22616/ESRD.2022.56.047478

Koçak, E., Ulucak, R., & Ulucak, Z. S. (2020). The impact of tourism developments on CO_2 emissions: An advanced panel data estimation. *Tourism Management Perspectives, 33*(2020), 100611.

Kusuma, H., Muafi, M., AJi, M. H., & Pamungkas, S. (2020). Information and communication technology adoption in small- and medium-sized enterprises: Demographic characteristics. *Journal of Asian Finance, Economics and Business, 7*(10), 969–980.

Li, J., & Tang, Y. (2010). CEO Hubris and firm risk taking in China: The moderating role of managerial discretion. *Academy of Management Journal, 53*(41), 45–68. https://doi.org/10.1111/1467-9442.00135

Marx, S., Flynn, S., & Kylanen, M. (2021). Digital transformation in tourism: Modes for continuing professional development in a virtual community of practice. *Project Leadership and Society, 2*(100034), 1–11.

Mazzone, D. M. (2014). *Digital or death: Digital transformation — The only choice for business to survive smash and conquer* (1st ed.). Smashbox Consulting.

Meggiolaro, S. (2018). Information and communication technologies use gender and mathematics achievement: Evidence from Italy. *Social Psychology of Education, 21*(2), 497–516. https://doi.org/10.1007/s11218-017-9425-7

Mhlanga, D., Ndhlovu, E., & Hofisi, C. (2021). Assessment of the 4IR challenges of radical innovation in service delivery in Africa. *Journal of Public Administration, 56*(4.1), 1002–1017.

Mudasir, N., Ghausee, S., & Stanikzai, H. U. R. (2020). A review of ecotourism and its economic, environmental and social impact. *Journal of Emerging Technologies and Innovative Research, 7*(8), 1739–1743.

Narayan, R., Gehlot, A., Singh, R., Akram, S. V., Priyadarshi, N., & Twala, B. (2022). Hospitality feedback System 4.0: Digitalization of feedback system with integration of industry 4.0 enabling technologies. *Sustainability, 14*, 12158.

Ndhlovu, E., & Dube, K. (2023). Challenges of radical technological transition in the restaurant industry within developing countries. *African Journal of Hospitality, Tourism and Leisure, 12*(1), 156–170.

Ndhlovu, E., & Mhlanga, D. (2023). The effects of COVID-19 on informal traders in undesignated spaces. *Emerging Science Journal, 7*(Special Issue), 95–104.

Nguyen, A. H., Ha, H. H., & Nguyen, S. L. (2020). Determinants of information technology audit quality: Evidence from Vietnam. *Journal of Asian Finance, Economics and Business, 7*(4), 41–50. https://doi.org/10.13106/jafeb.2020.vol7.no4.41

Okundaye, K., Fan, S. K., & Dwyer, R. J. (2019). Impact of information and communication technology in Nigerian small-to-medium-sized enterprises. *Journal of Economics, Finance and Administrative Science, 24*(47), 29–46. https://doi.org/10.1108/JEFAS-08-2018-0086

Opute, A. P., Irene, B. O., & Iwu, C. G. (2020). Tourism service and digital technologies: A value creation perspective. *African Journal of Hospitality, Tourism and Leisure, 9*(2), 1–18.

PwC. (2013). *Digital Transformation – der gro€ßte Wandel seit der Industriellen Revolution.* PricewaterhouseCoopers.

Ransbotham, S., Lurie, N. H., & Liu, H. (2019). Creation and consumption of Mobile word of mouth: How are Mobile reviews different? *Marketing Science, 38*, 773–792.

Rocha, Á., Correia, A. M., Adeli, H., Reis, L. P., & Teixeira, M. M. (2016). *New advances in information systems and technologies*. Springer.

Schallmo, D., Williams, C. A., & Boardman, L. (2017). Digital transformation of business models — Best practice, enablers, and roadmap. *International Journal of Innovation Management, 21*(8), 1–17. https://doi.org/10.1142/S136391961740014X

Shannon, C. (1948). A mathematical theory of communication. *Bell System Technical Journal, 27*(3), 379–423.

Shereni, N. C. (2022). *Tourism and sustainable development goals in Zimbabwe: Contribution by the hospitality sector*. (PhD Thesis) University of Johannesburg.

Shereni, N. C., & Chambwe, M. (2020). Hospitality Big Data analytics in developing countries. *Journal of Quality Assurance in Hospitality and Tourism, 21*(3). https://doi.org/10.1080/1528008X.2019.1672233

Shereni, N. C., Saarinen, J., & Rogerson, C. M. (2023). Sustainability drivers and challenges in the hospitality sector in Zimbabwe. *Tourism: An International Interdisciplinary Journal, 71*(3), 492–504.

Shimmura, T., Ichikari, R., Okuma, T., Ito, H., Okada, K., & Nonaka, T. (2020). Service robots introduction to a restaurant enhances both labour productivity and service quality. *Procedia CIRP, 88*, 589–594.

Solnet, D., Subramony, M., Ford, R. C., Golubovskaya, M., Kang, H. J., & Hancer, M. (2019). Leveraging human touch in service interactions: Lessons from hospitality. *Journal of Service Management, 30*(3), 392–409. https://doi.org/10.1108/josm-12-2018-0380

Swarbrooke, J. (2000). *Sustainable Tourism Management*. CABI Publishing.

Thrasso, A., Chebbi, H., & Uzunboylu, N. (2021). Postmodern approaches to business management and innovation for contextual adaptation: A review. *EuroMed Journal of Business, 16*(3), 261–273.

Troshani, I., Janssen, M., Lymer, A., & Lee, D. (2018). Digital transformation of business-to-government reporting: An institutional work perspective. *International Journal of Accounting Information Systems, 31*, 17–36.

Westerman, G., Calméjane, C., Bonnet, D., Ferraris, P., & McAfee, A. (2011). *Digital transformation: A roadmap for billion-dollar organizations* (pp. 1–68). MIT Center for Digital Business and Capgemini Consulting.

World Commission on Environment and Development. (1987). *Our common future. World Commission on environment and development*. Oxford University Press.

Xiang, Z., & Fesenmaier, D. (2017). Big data analytics, tourism design and smart tourism. In *Analytics in smart tourism design: Concepts and methods* (pp. 299–307). Springer.

Youssef, A. B., & Zeqiri, A. (2021). Hospitality Industry 4.0 and climate change. *Circular Economy and Sustainability*. https://doi.org/10.1007/s43615-021-00141-x

Zhang, J., & Shang, Y. (2022). The influence and mechanism of digital economy on the development of the tourism service trade—Analysis of the mediating effect of carbon emissions under the background of COP26. *Sustainability, 14*, 13414. https://doi.org/10.3390/su14201341

Part I
Conceptualisation and Review of Digital Technologies in Tourism

Chapter 2
A Review of Tourism and Hospitality Education and Training Using Multiple Intelligence Theory in the Digital Age of Learning in Sub-Saharan Africa

Tawanda Makuyana, Emmanuel Ndhlovu, and Kaitano Dube

2.1 Introduction and Background

Tourism and hospitality education has a background of having a disjoint in the course structure (setting) and training goals (Zhang & Zhou, 2021). Furthermore, off-the-job skills development needs to match on-the-job skills practice and grooming of talents within the tourism and hospitality value chain (Daniel et al., 2017). Stakeholders like educators, students, and industry-role players (employees and expectations of visitors included) in the value chain believe in synergies and input-output relationships between tourism and hospitality education and training and industry (Kumar et al., 2020). However, industry role players think that off-the-job skills development needs to address adequate human resources with occupation-based competencies, especially the young generation (Pani et al., 2015). However, the present chapter does not argue for on-the-job skills development. Still, a stance to bring a complementary strategic angle that can augment hospitality and tourism education and training (classroom-based learning to speak to the quality of services and hospitality and tourism work-based practices are missing in the tourism and hospitality education management and practices debates (Ghosh & Jhamb, 2022).

There are 88 universities in the 20 countries that constitute the Sub-Saharan region (The Times Higher Education, 2023). Brown et al. (2015) explored the perceptions of Generation Y with hospitality education from a career development perspective. Jackson and Wilton (2017) investigated perceived employability among undergraduates and the importance of career self-management, work experience, and individual characteristics. More research needs to be focused on the ability of educators to equip students with knowledge and skill sets in ways that uphold

---

T. Makuyana (✉) · E. Ndhlovu · K. Dube
Vaal University of Technology, Vanderbijlpark, South Africa
e-mail: kaitanod@vut.ac.za

E. Ndhlovu et al. (eds.), *Tourism and Hospitality for Sustainable Development*,
https://doi.org/10.1007/978-3-031-63069-9_2

workplace expectations. Such has implications for the value co-creation in tourism education (Padurean & Maggi, 2016).

Unfortunately, it is observed that in Sub-Saharan tourism and hospitality, education and training recruited most educators, especially in traditional universities, who have tourism expertise gained from only university degrees (qualifications). On the contrary, most of their counterparts in technical-skewed institutions like hotel schools and universities of technology acquired degrees and have work exposure in the tourism value chain (industry) (Makuyana & Saayman, 2018). It is vague on how the knowledge and skills held by educators can be transferred in this digital era, especially towards the inclusion of every student with different learning abilities. However, the literature indicates that the extent of students' preparedness to join the labour market is a by-product of educators' knowledge and skill set, which would have been interpreted into curricula and learning outcomes and employability thereof (O'Leary, 2017). Hence, qualified students reflect both the effects of teaching and learning processes and individual learning regardless of digital teaching and learning enhancement (Abelha et al., 2020). Hence tourism and hospitality industry role players continuously blame the educators and off-the-job skills development system for failing to prepare graduates with competencies needed in the industries (Nhuta et al., 2015). However, it is expected that chefs and bakers in most technical tourism and hospitality institutions have even acquired teaching qualifications to be equipped with classroom management (pedagogies), among other teaching competencies (Makuyana, 2014). In contrast, traditional universities encourage educators to study for higher degrees like doctorates for knowledge development at the expense of knowledge dissemination to students in manners that uphold learning among diverse students.

Despite these disparities, no reviews or research focus on the relationship between educators' qualifications and employability of human capital products of higher education and training in tourism and hospitality in the Sub-Saharan region, especially from when the fourth industrial revolution began until the post-health pandemic. The present chapter is essential to the tourism and hospitality stakeholders in providing insights towards addressing classroom-based learning using similar approaches used when the workforce executes work tasks in the world of work. It, therefore, identified gaps in tourism and hospitality education and training literature on teaching and learning since the inception of the fourth industrial revolution, the health pandemic, and resilience pathways in the post-health pandemic era. In that lens, the chapter would propose and encourage lecturers in tourism and hospitality higher education to adopt progressive teaching approaches to accommodate the varied 'abilities' of students and enable these students to excel in their learning.

On the one hand, tourism and hospitality tertiary education are located in an industry space where disruptive technologies, platforms, and principles of stewardship have contributed to service delivery (Buhalis et al., 2019), and such affect learning work-related skills in the education system (Sheldon et al., 2016). On the other hand, the tourism and hospitality value chain is adopting technological advancements like using robots and artificial intelligence in business operations (Nayyar et al., 2018; Ndhlovu & Dube, 2023b, c; Tung & Au, 2018). Therefore, it is essential to bring the digital age of learning and means to enable learning among students with different learning abilities as a layer into the tourism and hospitality education and training debate.

The present chapter seeks to bring a combination of digitalised and non-digitality technologies to nurture the need to adopt the Multiple Intelligence Theory (MIT) principles into tourism and hospitality education and training because there is limited use of this theory in the field mentioned above (Barrington, 2004; Subia et al., 2022). The discourse is supported by Sener and Çokçaliskan (2018), who found a positive relationship between multiple intelligences and learning styles. Therefore, following Stanford (2003), MIT enables teaching (lecturing) to effectively understand and address students in a class while preparing them for the workspace. MIT upholds incorporating an element of creativity into the student-centric instructional design (Stanford, 2003). The instructional design can enable conducting consistent activities in eight different forms to reach students with various abilities and learning styles, as Howard Gardner proposed in 1983 in his Multiple Intelligence Theory (MIT) (Armstrong, 2009).

According to Adcock (2014) and Armstrong (2009), MIT indicates that every individual has an intelligence profile composed of eight intelligences: visual/spatial, verbal/linguistic, musical/rhythmic, logical/mathematical, bodily/kinaesthetic, interpersonal, intrapersonal, and naturalistic. This implies that tourism and hospitality students fall into categories where they are more robust in certain types or weaker in some. As Adcock (2014) and Armstrong (2009) allude, visual/spatial intelligence is composed of the ability to manipulate and create mental images and remember facts by visualising. Linguistic intelligence involves mastery of language; learners are sensitive to the meaning of language and words (Adcock, 2014; Armstrong, 2009). Musical/rhythmic intelligence involves recognising non-verbal sounds in the environment and being sensitive to pitch, melody, tone, and rhythm (Adcock, 2014; Armstrong, 2009). Logical/Mathematical Intelligence consists of noticing numerical or analytical patterns (Adcock, 2014; Armstrong, 2009). Bodily/kinaesthetic intelligence is the ability to express oneself physically (Adcock, 2014; Armstrong, 2009). Interpersonal intelligence includes understanding and relating well to others (Adcock, 2014; Armstrong, 2009). Intrapersonal intelligence involves identifying and understanding one's feelings and emotions (Adcock, 2014; Armstrong, 2009). Naturalistic intelligence involves connecting to the natural environment and is suitable for identification and observation (Adcock, 2014; Armstrong, 2009). The intelligence highlighted by Adcock (2014) and Armstrong (2009) are critical competencies and attributes for one to be a thriving tourism and hospitality professional, whether in academics or the industry in the sector and sub-sectors. It is observed that more needs to be done both in the tourism and hospitality research community and education and training because it poses implications for the tourism and hospitality sector's skills development and value chain (Nhuta et al., 2015).

Resilience from the health crisis exacerbated the current debate on the feasibility of fully adopting digitalisation in tourism and hospitality education and training amidst the controversy of tertiary education can equip students with the industry-expected competencies (Ndhlovu & Dube, 2023a; Reichenberger & Raymond,

2021; Rivera et al., 2021; Seo & Kim, 2021). The debates-tourism and hospitality education and training versus industry-needed human capital followed various dimensions during the health pandemic (Qiu et al., 2021). However, according to Shen et al. (2022), the fundamentals of tourism and hospitality education and training are centred on aspects such as the tourism and hospitality content co-creation, student learning process, co-creation of learning environment, assessment and engagement of learners, teaching practices, and accessibility of learning among various learners with different socio-economic background and geographical locations, which were related to augmented reality and virtual reality.

The questions that seem to be lingering among stakeholders, in this case, educators is: Based on the multi-faceted tourism and hospitality sector, can technical institutions (universities, polytechnics, hotel schools) and liberal/traditional universities that offer different curricula adopt the same digital teaching and learning approaches in the sub-Saharan region? Is there any institution using MIT in tourism and hospitality education and training? Is there evidence of success, challenges, and opportunities in using this model for effective pedagogy, teaching practices (including their implementation in the classroom and impact on student learning), and technological enhancements to teaching in tourism and hospitality education and training? The chapter presents the methodology and material used for the study, the findings, and the implications that position the gaps and impacts as shown in the extant literature. The conclusion upholds the proposition for future research agenda. As informed by the findings, it postulates a theoretical framework for various tourism and hospitality education and training in the sub-Sahara region.

2.2 Methods and Materials

Based on geo-political contexts that influence growth in tourism and hospitality education and training, the study used a scoping literature review to explore the issue under investigation. It is critical because of the ever-on-going argument between stakeholders in tourism and hospitality education and training and the industry value chain. Data was generated from Google Scholar (handles the majority of predatory and nonpredatory research-based-output published in the public domain) and SCOPUS (handles non-predatory-based research output published in the public domain) because they are one of the largest international databases that have Sub-Saharan tourism and hospitality education, and training research hat upholds the digital age of learning concept. It was deliberately done to establish the scoping literature review using all available existing and accessible extant literature. The scoping literature review considered research between 2016 and 2023 to have a comparative perspective of the pre-health pandemic, during, and post-crisis resilient pathways within the fourth industrial revolution influencing tourism and hospitality education and training.

2.2.1 *Data Gathering*

The first author logged into SCOPUS using institutional details and searched using the article title, abstract, and keywords like tourism AND hospitality AND education AND training AND technology (as synonyms to the digital age of learning). The search was filtered from 2016 to 2023 to get recent publications since the inception of the fourth industrial revolution. Article title, abstract, and keywords were the focus of the search. Only 30 documents came out. Then, the same author searched tourism AND hospitality AND education AND training AND multiple AND intelligence AND theory AND digital AND learning. Only one document came out. Leading to the widening of the search using Google Scholar.

The first author used Google Scholar to search within keywords, article title, and abstract. The first step of the screening process to initiate the electronic searches within Google Scholar was done using the phrases: tourism and hospitality education and training using multiple intelligence theory; tourism and hospitality education and training using multiple intelligence theory in the digital age of learning in the sub-Sahara region. Results came with 7880 results.

The researcher manually opened and read each abstract for the research documents in the electronic search conducted on Google Scholar as the first step in screening. All abstracts that did not address tourism and hospitality education and training (technical education) and technology (including all forms of non-Internet and Internet-based digital technologies) were excluded. In addition, all abstracts that did not address tourism and hospitality education and training using Multiple Intelligence Theory in a digital learning age were excluded. Studies that were not on open access were excluded, too. Abstracts that scoped tourism and hospitality education/training using Multiple Intelligence Theory in the digital era/age of learning as a single study were included. Therefore, the chapter only had open-access research documents. Two folders were created to store Google Scholar and Scopus data, respectively.

2.2.2 *Data Analysis*

VOS Viewer was used to analyse co-occurrences of keywords using the full counting method to produce a network visualisation for documents from SCOPUS only because Google Scholar still needs to be incorporated into VOSviewer-compatible databases. Out of 151 items (selected keywords), only 81 items were connected. However, a content analysis approach was used in analysing all gathered data. The data were then examined from multiple intelligence theory lenses while nurturing Vergnes et al.'s (2010: 771) scientific way of synthesising a plethora of information after exhaustively searching and objectively analysing (reduction, exploration of text). Within the process, the authors assessed the application of MIT in tourism and hospitality education and training. Full-text reading ascertained the level of

Table 2.1 Electronic research work—search criteria

The current literature research and selection criteria		
Aim: To scope the extent and how MIT is being used in tourism and hospitality education and training in the digital era of learning.	*Inclusion*	*Exclusion*
	Using the articles' keywords	Articles that do not align with keywords
	Peer-reviewed full-text in English	Articles that are not peer-reviewed and non-English
	Peer-reviewed published articles on MIT	Non-peer-reviewed reports and handbooks
	Academic work on tourism and hospitality education and training	Non-academic work
	Research focus aligned with the following: Can technical institutions (universities, polytechnics, hotel schools) and liberal/traditional universities that offer different curricula adopt the same digital teaching and learning approaches in the sub-Saharan region? Is there any institution using MIT in tourism and hospitality education and training? Is there evidence of success, challenges, and opportunities in using this model for effective pedagogy, teaching practices (including their implementation in the classroom and impact on student learning), and technological enhancements to teaching in tourism and hospitality education and training?	

Source: authors' compilation

alignment with the purpose of the review. All materials which did not address keywords were discarded (see Table 2.1).

Table 2.1 lists the objective guiding criteria used in the electronic literature search and establishes categories to develop a structured report after the review process.

The content-thematic analysis was adopted as full text in the third reading of the selected documents. The second and third authors did the fourth and fifth readings to affirm the establishment of the critical categories that fostered the presentation of findings. In this context, a five-step content-thematic analysis was used to analyse the data: logging, anecdotes, vignettes, data coding, and thematic networking (Creswell, 2014). It enabled identifying, analysing, and reporting patterns as themes within a text (Snyder et al., 2019).

From this, vignettes were developed by both narrative and investigations fostered in higher levels of interpretation, clarity, and rigour (credibility) on the gathered tourism and hospitality education within the digital age of learning (Creswell & Poth, 2016). Hence, the in-depth description of environments, methodologies, participants, and areas of focus set by each article fostered the rigour/credibility of the studies as rooted in a grounded theory lens (Khan, 2014). Data coding based on MIT elements was set, and the codes were clustered into themes, respectively.

The themes are (i) digital teaching and learning approaches, (ii) usage of MIT in tourism and hospitality education and training, (iii) evidence of success in using MIT for effective pedagogy, teaching practice, and technological enhancement to teaching in tourism and hospitality education and training, (iv) evidence of challenges in using MIT for effective pedagogy, teaching practice, and technological enhancement to teaching in tourism and hospitality education and training, (v) evidence of opportunities in using MIT for effective pedagogy, teaching practice, and technological enhancement to teaching in tourism and hospitality education and training.

Rigour was ascertained because the peer-reviewed scientific-published journal articles and theses went through blind peer-review by at least two reviewers or more and examiners for the latter. It is augmented by the current scientific review process set by the present authors, thereby enhancing the rigour and trustworthiness of the gathered data. In addition, the trustworthiness of this paper is centred on ethical considerations. The first author went through ethics training, which complements his research experience from PhD studies that focused on tourism human capital development/education and training. In addition, the second author-verified quality based on his vast research experience, and the third as a mentor in post-PhD research endeavours. It ensures the quality of the review process for this current article.

2.3 Results

The results are presented separately as follows:

SCOPUS-based data input, which VOSviewer analysed, indicates that hospitality education has one occurrence but with 11 links. In contrast, technology has 2 occurrences and 19 links to teaching, covid, hospitality education, education, labour market, service activities, and cultural innovation (see Fig. 2.1).

According to network visualisation in Fig. 2.1, teaching has two occurrences with 26 links to education, e-learning, students, coursework, international tourism, learning approaches, and computer-aided instructions. Tourism education has three occurrences with ten links: tourism, higher education, distance education, e-learning, training, and retraining (see Fig. 2.1). Students have two occurrences with 22 links that entail virtual reality, future employee training, e-learning, communication skills, and teaching. Hospitality has three occurrences with 15 links: tourism, mobile learning, technology-enhanced learning, and ubiquitous learning. Tourism has five occurrences and 28 links that entail development, hospitality, managers, education, educational programs, educational technologies, staff training, and vocational education.

From an overlaying visualisation (see Fig. 2.1), vocational education, educational programs, and staff training were observed to have received research attention around 2016. Manager and tourism have surfaced from 2017 to 2019, while tourism

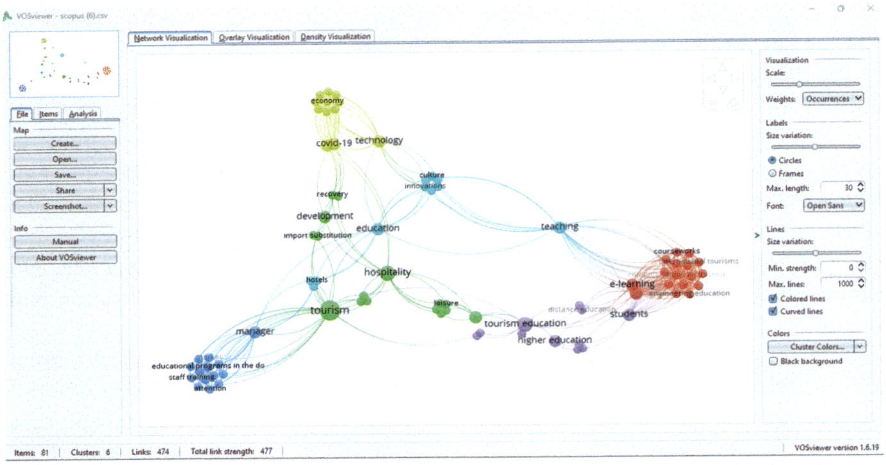

Fig. 2.1 Network visualisation. (Source: VOS viewer)

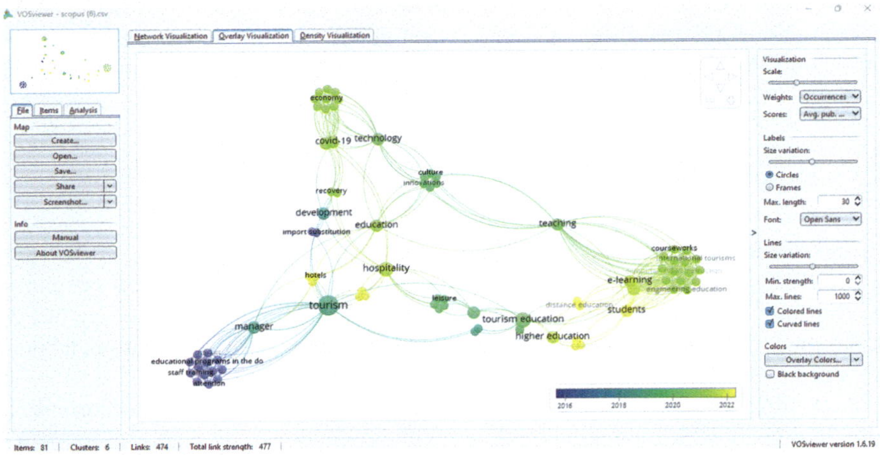

Fig. 2.2 Overlaying visualisation. (Source: VOS viewer)

education and teaching, coursework, learning performance, and computer-aided instruction and technology gained attention in research around 2020. E-learning, hospitality, students, higher education, distance education, and online learning gained popularity through research in post-2021 (Fig. 2.2).

Based on Scopus-based analysis, in 2016, two research documents were published that are accessible. A decrease is observed in 2017 and 2018, then two documents are available in 2019, and 4 in 2020, followed by 9 in 202. A decline is evident as 2022 had only seven documents, totalling 30 in the highlighted period (see Fig. 2.3).

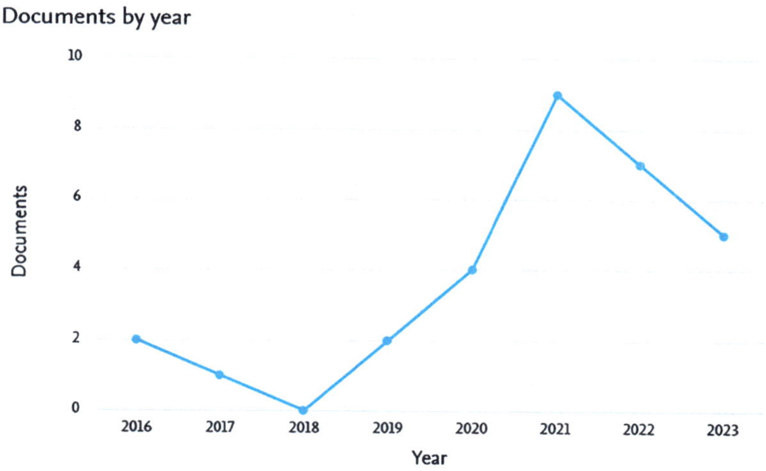

Fig. 2.3 Documents by year. (Source: Scopus database)

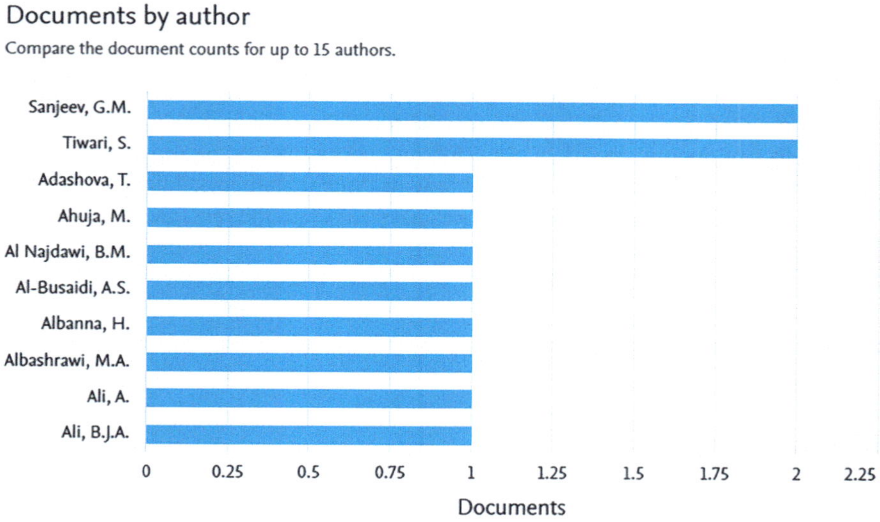

Fig. 2.4 Documents by author. (Source: Scopus database)

The results indicate that the affiliations among authors and these institutions are mainly located in the Global North. At the same time, the Global South, especially researchers based in Africa, is limited.

It implies that the tourism and hospitality education and training sub-Saharan region still needs to focus on education and training using MIT. In addition, it is observed that only two authors have two documents, and the rest have one, respectively (see Fig. 2.4).

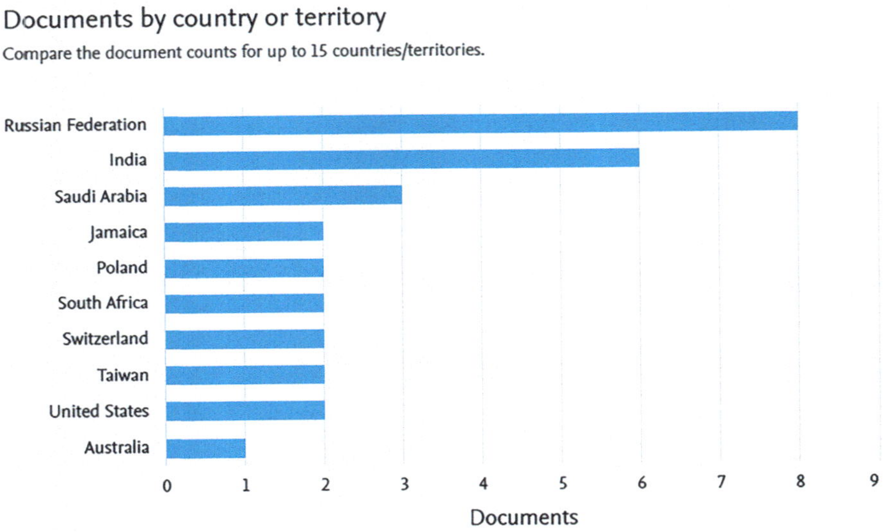

Fig. 2.5 Document by country/territory. (Source: Scopus database)

Most authors are based outside Africa, as the results reflect documents by country/territory. Only South Africa represents the sub-Saharan region on the topic under study, with two documents. Thus, the Russian Federation has the most (eight), followed by India with six and Saudi Arabia with three documents. Jamaica, Poland, Switzerland, Taiwan, and the United States have two, and the rest have one, respectively (see Fig. 2.5).

The results in Fig. 2.6 reflect the type of documents: (i) article constitutes the majority with above 63%, (ii) conference papers with 16.7% and conference reviews had the least.

2.4 Discussion

Based on the results of Scopus and VOSviewer, the multi-faceted tourism and hospitality sector that gets labour from technical institutions (universities, polytechnics, hotel schools) and liberal/traditional universities (labour market) cannot adopt the same digital teaching and learning approaches in the sub-Saharan region. It is due to the following: (i) interest in tourism and hospitality education and training among researchers/educators, (ii) the objectives of the various curricula, (iii) the relationship between training, the labour market, and perceived vocational training needed, (iv) the feasibility of computer-aided instruction, e-learning, and online learning, (v) tourism and hospitality education and training environment/space. Little needs to be mentioned about using MIT as augmented by digital-based technologies for

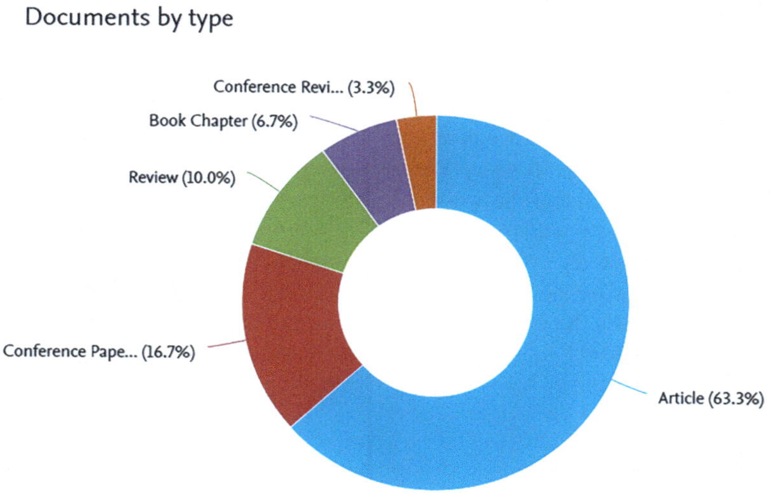

Fig. 2.6 Documents by type. (Source: Scopus database)

teaching and learning purposes, especially in Sub-Saharan Africa and other regions and continents in the Global North. (vi) Exploring the relationship between educators' knowledge and skill sets, digitally augmented teaching and learning processes, various learning abilities, and human capital readiness and employability in the industry value chain.

Results show no higher education and training institution that uses principles derived from MIT in tourism and hospitality education and training. It implies that there is yet to be evidence of success, challenges, and opportunities in using MIT for effective pedagogy, teaching practices (including their implementation in the classroom and impact on student learning), and technological enhancements to teaching in tourism and hospitality education and training. Despite that, there is the adoption of digital technologies in tourism and hospitality education and training in various dimensions like e-commerce, hotel employee training using virtual reality (Fiala et al., 2022), e-learning in tourism education (Kalbaska & Cantoni, 2022), modernisation of higher tourism and hospitality education, competencies models (Vogel et al., 2022). These layers have not considered various bits of intelligence possessed by people with and without disabilities who seek to pursue tourism and hospitality professional occupations from accessible tourism.

Results from Google Scholar brought results that had research documents in education, among other fields. Therefore, out of 7880, only eight were related to the topic under study. The documents included: (i) teaching cross-cultural management: A flipped classroom approach using films (Desai et al., 2018), (ii) The digital revolution: Impact on tourism education by Cinar (2020), (iii) Human Resource Skill Adjustment in Service Sector: Predicting Dynamic Capability in Post COVID-19 Work Environment (Zayed et al., 2022), (iv) Transitioning vocational education and training in Africa: a social skills ecosystem perspective, VET Africa

4.0 Collective (Spours, 2023a, b), (v) Blended Learning in the Vocational Education and Training System in Tanzania: Understanding Vocational Educators' Perceptions (Muchum et al., 2016), (vi) Vocational Education and Training: Basics for teaching and research in Vocational Education and Training at Universities (Duncan et al., 2017), (vii) Developing the Effective Digital Human Resource Management What Experts Say for Tourism Industry Preparations (Sitaniapessy et al., 2021), (viii) Modelling the relationship between higher education service quality, student engagement, attachment, satisfaction, and loyalty: a case of a Malawian public university (Kankhuni et al., 2023). The Google Scholar database upholds limited research that addresses the topic under study. However, extant literature shows a build-up that supports the current argument that tourism and hospitality education and training can use MIT to augment the digital learning age.

2.4.1 *Implications*

Based on the multi-faceted tourism and hospitality sector that is continuously acknowledging and adapting teaching and learning in support of SDGs that seek to leave no one behind and diversity in learning abilities using examples like e-learning and computer-aided instructions. These take cognisance that students choose tourism and hospitality programs as the last options after failing to secure marks required for other programs for professions perceived as better in society. With this in mind, it is observed in extant literature that, as digitalisation facilitates tourism and hospitality education and training, such has yet to consider diversities in learning abilities to the extent of adopting MIT in sub-Saharan Africa. On the contrary, the sub-Saharan region has previously marginalised population groups with various learning abilities, like poor people, people with learning disabilities, and even the rich ones who would not have proved to have learning abilities to do other professions and would have decided to pursue tourism and hospitality due to its limited barriers to entry into the tourism and hospitality programs. As reflected in the findings, technical institutions (universities, polytechnics, hotel schools) and liberal/traditional universities that offer different curricula cannot adopt the same digital teaching and learning approaches in the sub-Saharan region. However, MIT is feasible to embed in that technical institutions and liberal universities' digital teaching and learning seek to address psychomotor, cognitive, and affective domains.

The findings indicated that technical universities, polytechnics, hotel schools, and liberal universities that offer tourism and hospitality programs adopt various digital-based teaching and learning as driven by a variety of contexts, like educators' digital competencies, accessibility of digital learning platforms among learners based on socio-economic backgrounds and location, resources base to support the learning needs among various learners, diversity in students' learning process, co-creation of learning environment, assessment, engagement and motivation of learners and structural characteristics that distinguish, hospitality and tourism as believed by Singal (2015).

The findings show that institutions offering tourism and hospitality have yet to consider using MIT in education and training. Nhuta et al. (2015) indicate that countries like Zimbabwe, among other southern African states, have tourism and hospitality institutions that hardly prepare their students using approaches and competencies that resemble industrial/occupations in the value chain. The gap has led private organisations to prefer in-house training to off-the-job classroom-based learning. It is because industry role players believe in using and targeting different intelligences that enable learning abilities to equip a tourism and hospitality professional with practical knowledge and skill sets. The gap can be closed by a willingness to invest in tourism and hospitality education and training—industry/ work-resembling settings guided by industry-role involvement in curricula development processes.

Based on the findings, there is scant evidence of success, challenges, and opportunities in using MIT for effective pedagogy, teaching practices (in the classroom and impact on student learning), and technological enhancements to teaching in tourism and hospitality education and training. It implies that knowledge co-creation and skill set impartation should involve occupation-based standards, of which states in the sub-Saharan region have a more industry-based association to further owners' interests. At the same time, there are limited non-academic professional associations/boards, if not only Tour Guiding, Chefs, and Bakers, to regulate professional training, work decent, career image, and standards of training and practices. The other occupational jobs, like housekeeping and travel consultants in the tourism and hospitality sub-sectors, need to be more structured. Hence, increasing vulnerability in knowledge and skills development, especially in content co-creation, understanding various student learning processes, co-creation of a conducive learning environment for students with different learning abilities, assessment and engagement of learners, teaching practices, and accessibility of learning among various learners with diverse socio-economic background and geographical locations.

2.5 Conclusion

In conclusion, the chapter acknowledges the multi-faceted tourism and hospitality sector that has technical institutions (universities, polytechnics, hotel schools) and liberal/traditional universities that offer different curricula as means to facilitate the preparation of workforce/human resources to be ready for the world of work using digital teaching and learning approaches in the sub-Saharan region. The education and training fraternity can embed MIT in its various digital teaching and learning approaches to accommodate multiple learning needs among learners in sub-Saharan Africa.

Notably, there has yet to be an institution using MIT in the Sub-Saharan region's tourism and hospitality education and training. Therefore, there is scant evidence of success, challenges, and opportunities in using MIT for effective pedagogy, teaching practices (including their implementation in the classroom and impact on student learning), and technological enhancements to teaching in tourism and hospitality education and training. Therefore, the chapter recommends training the trainers (in-service) at MIT in tourism and hospitality education and training.

Furthermore, the chapter acknowledges the tourism and hospitality-specific areas, contexts, and differences among tertiary institutions. It, therefore, concludes by deducing and postulating theoretical models for effective pedagogy, teaching practices (including their implementation in the classroom and their impact on student learning), and technological enhancements to teaching in tourism and hospitality education and training for the various tertiary institutions in the sub-Saharan region (see Fig. 2.7).

The chapter proposed that hospitality education and training can augment teaching approaches that uphold learning among different learning abilities of students with digital learning approaches, as shown in Fig. 2.7.

Figure 2.7 illustrates that the tourism and hospitality curricula are designed, and content is generated by educators based on the aims and outcomes of the curricula. The emphasis given by the chapter is on the style of delivering the content to enhance student learning processes. Thus, the classroom environment is considered in both technical and liberal education and training institutions. Based on occupational-based competencies, the teaching styles/approaches can embed MIT principles to prepare students with different learning abilities for the world of work.

The study's limitations are using only two international databases on specific theories. Hence, a broader picture can be obtained if all databases are explored to complement empirical studies in the sub-Saharan region. The chapter suggests that future research on digital-based tourism and hospitality teaching and learning consider the following areas: (i) learning of persons with intellectual disabilities, (ii) training educators in teaching and classroom management (pedagogy), (iii) sub-sector-specific teaching (education and training) enhancements for the learning

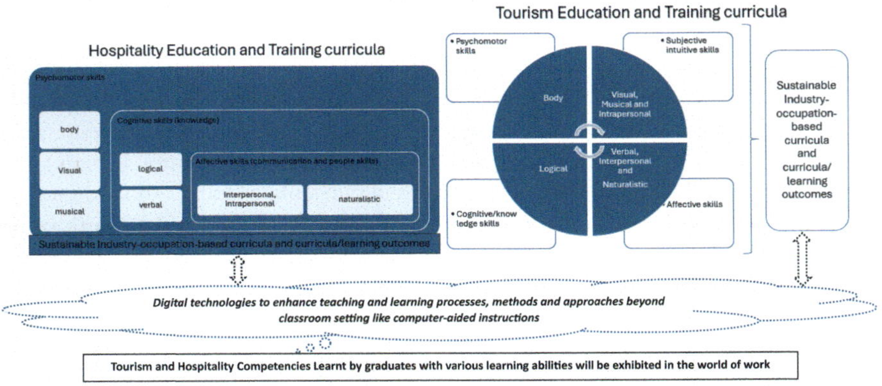

Fig. 2.7 Theoretical frameworks for tourism and hospitality education and training using MIT. (Source: Deducted from the findings)

processes for technical- and liberal university-based curricula, (iv) linkage of career growth, workplace-based knowledge and skill sets, and formal qualifications, (v) effectiveness of MIT in tourism and hospitality education and training.

References

Abelha, M., Fernandes, S., Mesquita, D., Seabra, F., & Ferreira-Oliveira, A. T. (2020). Graduate employability and competence development in higher education—A systematic literature review using PRISMA. *Sustainability, 12*(15), 5900.

Adcock, P. K. (2014). The longevity of multiple intelligence theory in education. *Delta Kappa Gamma Bulletin, 80*(4), 50.

Armstrong, T. (2009). *Multiple intelligences in the classroom*. Ascd.

Barrington, E. (2004). Teaching to student diversity in higher education: How multiple intelligence theory can help. *Teaching in Higher Education, 9*(4), 421–434.

Brown, E. A., Thomas, N. J., & Bosselman, R. H. (2015). Are they leaving or staying: A qualitative analysis of turnover issues for Generation Y hospitality employees with a hospitality education. *International Journal of Hospitality Management, 46*, 130–137.

Buhalis, D., Harwood, T., Bogicevic, V., Viglia, G., Beldona, S., & Hofacker, C. (2019). Technological disruptions in services: Lessons from tourism and hospitality. *Journal of Service Management, 30*(4), 484–506.

Çinar, K. (2020). The digital revolution: Impact on tourism education. *Journal of Tourism & Gastronomy Studies, 8*(4), 2417–2443.

Creswell, J. W. (2014). *A concise introduction to mixed methods research*. SAGE Publications.

Creswell, J. W., & Poth, C. N. (2016). *Qualitative inquiry and research design: Choosing among five approaches*. Sage Publications.

Daniel, A. D., Costa, R. A., Pita, M., & Costa, C. (2017). Tourism education: What about entrepreneurial skills? *Journal of Hospitality and Tourism Management, 30*, 65–72.

Desai, S. V., Jabeen, S. S., Abdul, W. K., & Rao, S. A. (2018). Teaching cross-cultural management: A flipped classroom approach using films. *The International Journal of Management Education, 16*(3), 405–431.

Duncan, K., Downing, J., D'Oliveira Singo, B., Papier, J., Hartmann, M. D., Ogwo, B. A., et al. (2017). Session 1 - vocational education and training: Basics for teaching and research in vocational education and training at universities. In G. Haseloff, F. Eicker, & B. Lennartz (Eds.), *Vocational education and training in sub-Saharan Africa: Current situation and development* (pp. 39–133). W. Bertelsmann Verlag. https://doi.org/10.3278/6004570w039

Fiala, J., Hán, J., Husák, J., Chadt, K., Chalupa, Š., Jenčková, J., et al. (2022, April). More than virtual reality: Tools, methods, and approaches to hotel employee training. In *The International Research & Innovation Forum* (pp. 435–443). Springer International Publishing.

Ghosh, P., & Jhamb, D. (2022). The role of Stakeholders in Enhancing Service Quality in Hospitality Education: An Application of Stakeholder Theory. Revista Turismo & Desenvolvimento (RT&D)/Journal of Tourism & Development, (38).

Jackson, D., & Wilton, N. (2017). Perceived employability among undergraduates and the importance of career self-management, work experience, and individual characteristics. *Higher Education Research & Development, 36*(4), 747–762.

Kalbaska, N., & Cantoni, L. (2022). E-Learning in tourism education. In *Handbook of e-Tourism* (pp. 1667–1686). Springer International Publishing.

Kankhuni, Z., Ngwira, C., Sepula, M. B., & Kapute, F. (2023). Modeling the relationship between higher education service quality, student engagement, attachment, satisfaction, and loyalty: A case of a Malawian public university. *Journal of Teaching in Travel & Tourism, 23*(2), 229–252.

Khan, S. N. (2014). Qualitative research method: Grounded theory. *International Journal of Business & Management, 9*(11), 224–233.
Kumar, G., Singh, G., Bhatnagar, V., Gupta, R., & Upadhyay, S. K. (2020). Outcome of online teaching-learning over traditional education during covid-19 pandemic. *International Journal, 9*(5), 77014–7711.
Machumu, H. J., Zhu, C., & Sesabo, J. K. (2016). Blended learning in the vocational education and training system in Tanzania: Understanding vocational educators' perceptions. *International Journal of Multicultural and Multireligious Understanding, 3*(2), 30–45.
Makuyana, T. (2014). *A comparative analysis of Zimbabwe's tourism curriculum needs and industrial manpower needs*. Masters thesis/dissertation, Midlands State University. Gweru. Zimbabwe.
Makuyana, T., & Saayman, M. (2018). *Transforming technical vocational hospitality and tourism training in developing economies: Case of Higher Examination Council of Zimbabwe*. Conference Proceedings: 'TVETising and STEMitising through Research and Innovation for Socio-Economic Transformation 8th and 9th March 2018 at the Troutbeck Resort, Nyanga, Zimbabwe. ISBN 978-0-7974-8922-6, EAN-9780797489226. pp. 39–63.
Nayyar, A., Mahapatra, B., Le, D., & Suseendran, G. (2018). Virtual Reality (VR) & Augmented Reality (AR) technologies for tourism and hospitality industry. *International Journal of Engineering & Technology, 7*(2.21), 156–160.
Ndhlovu, E., & Dube, K. (2023a). Impact of COVID-19 on tourism in South Africa: Building resilience and prospects for recovery. In K. Dube, G. Nhamo, & M. P. Swart (Eds.), *COVID-19, tourist destinations and prospects for recovery volume one: A global perspective* (pp. 81–100). Springer Nature.
Ndhlovu, E., & Dube, K. (2023b). Challenges of radical technological transition in the restaurant industry within developing countries. *African Journal of Hospitality, Tourism and Leisure, 12*(1), 156–170.
Ndhlovu, E., & Dube, K. (2023c). Restaurants industry recovery challenges from COVID-19 amidst the Russo-Ukrainian War. In K. Dube, G. Nhamo, & M. P. Swart (Eds.), *COVID-19, tourist destinations and prospects for recovery volume one: A global perspective* (pp. 337–356). Springer Nature.
Nhuta, S., Makuyana, T., Makoni, R., & Chauke, A. (2015). The disparity between the tourism education/training and the needs of the tourism and hospitality industry manpower needs. *Journal of Educational Policy and Entrepreneurial Research, 2*(5), 65–83.
O'Leary, S. (2017). Graduates' experiences of, and attitudes towards, the inclusion of employability-related support in undergraduate degree programmes; trends and variations by subject discipline and gender. *Journal of Education and Work, 30*(1), 84–105.
Padurean, L., & Maggi, R. (2016). TEFI values in tourism education: A comparative analysis. In *The tourism education futures initiative* (pp. 135–148). Routledge.
Pani, A. K., Srimannarayana, M., & Premarajan, R. K. (2015). e-Learning: Challenges and Solutions â€"A Case Study. International Journal of Learning, Teaching and Educational Research, 13(4).
Qiu, H., Li, Q., & Li, C. (2021). How technology facilitates tourism education in COVID-19: Case study of Nankai University. *Journal of Hospitality, Leisure, Sport & Tourism Education, 29*, 100288.
Reichenberger, I., & Raymond, E. M. (2021). Students' perceptions of tertiary tourism education and careers during the Covid-19 pandemic. *Journal of Teaching in Travel & Tourism, 21*(4), 380–402.
Rivera, M., Shapoval, V., & Medeiros, M. (2021). The relationship between career adaptability, hope, resilience, and life satisfaction for hospitality students in times of COVID-19. *Journal of Hospitality, Leisure, Sport & Tourism Education, 29*, 100344.
Sener, S., & Çokçaliskan, A. (2018). An investigation between multiple intelligences and learning styles. *Journal of Education and Training Studies, 6*(2), 125–132.
Seo, S., & Kim, H. J. (2021). How COVID-19 influences hospitality and tourism education: Challenges, opportunities, and new directions. *Journal of Hospitality & Tourism Education, 33*(3), 147–147.

Sheldon, P. J., Fesenmaier, D. R., & Tribe, J. (2016). The tourism education futures initiative (TEFI): Activating change in tourism education. In *The tourism education futures initiative* (pp. 14–35). Routledge.

Shen, S., Xu, K., Sotiriadis, M., & Wang, Y. (2022). Exploring the factors influencing the adoption and usage of Augmented Reality and Virtual Reality applications in tourism education within the context of the COVID-19 pandemic. *Journal of Hospitality, Leisure, Sport & Tourism Education, 30*, 100373.

Singal, M. (2015). How is the hospitality and tourism industry different? An empirical test of some structural characteristics. *International Journal of Hospitality Management, 47*, 116–119.

Sitaniapessy, R. H., Boediman, S. F., & Yulianti, E. B. (2021). Developing the effective digital human resource management what experts say for tourism industry preparations. *International Journal of Business, Economics and Management, 4*(20), 407–416.

Snyder, K. E., Fong, C. J., Painter, J. K., Pittard, C. M., Barr, S. M., & Patall, E. A. (2019). Interventions for academically underachieving students: A systematic review and meta-analysis. *Educational Research Review, 28*, 100294.

Spours, K. (2023a). Transitioning vocational education and training in Africa: A social skills eco-system perspective, VET Africa 4.0 collective. *Journal of Vocational Education & Training*. https://doi.org/10.1080/13636820.2023.2230037

Spours, K. (2023b). *Transitioning vocational education and training in Africa: A social skills ecosystem perspective, VET Africa 4.0 collective*. Bristol University Press.

Stanford, P. (2003). Multiple intelligence for every classroom. *Intervention in School and Clinic, 39*(2), 80–85.

Subia, G. S., Tangonan, M. E., Ricio, N. M. S., Pascual, G. G., Nuñez, R. N., & Mones, E. C. (2022). Understanding personality types and multiple intelligences of hospitality and tourism management graduating students: Inputs to skills enhancement and employability. *Open Journal of Business and Management, 10*(2), 606–613.

The Times Higher Education Sub-Saharan Africa University Rankings. (2023). [Online] https://www.timeshighereducation.com/sub-saharan-africa-university-rankings

Tung, V. W. S., & Au, N. (2018). Exploring customer experiences with robotics in hospitality. *International Journal of Contemporary Hospitality Management, 30*(7), 2680–2697.

Vergnes, J. N., Marchal-Sixou, C., Nabet, C., Maret, D., & Hamel, O. (2010). Ethics in systematic reviews. *Journal of Medical Ethics, 36*(12), 771–774.

Vogel, D., Lisboa Sohn, A. P., & de Santucci Oliveira, B. (2022). Analysis of scientific production on competencies models in Culinary Arts Education on EBSCO, web of science, and Scopus databases. *Journal of Culinary Science & Technology, 22*, 1–14.

Zayed, N. M., Edeh, F. O., Darwish, S., Islam, K. M. A., Kryshtal, H., Nitsenko, V., & Stanislavyk, O. (2022). Human resource skill adjustment in service sector: Predicting dynamic capability in post COVID-19 work environment. *Journal of Risk and Financial Management., 15*(9), 402.

Zhang, C., & Zhou, X. (2021). Curriculum settings and comparisons. In *Hospitality and tourism education in China: Development, issues, and challenges* (pp. 154–175). Routledge.

Part II
Digital Innovation in Tourism and Hospitality

Chapter 3
Digital Innovation Adoption in South African National Parks, Hotels, and Airports

Peter Chihwai

3.1 Introduction

The COVID-19 pandemic accelerated digitization in the tourism sector globally. Tourism departments and ministries including individual tourism organizations conceptualized, developed, and implemented digital innovations to circumvent the dire situation created by the disease and sustain the various tourism activities. The application of technology started years back but as lip service and companies were not taking serious action such as in the empirical findings in Hong Kong hotels' strategic decision-making processes (Law & Jogaratnam, 2005). Highlighting further, the underutilization of information technology globally (Law & Jogaratnam, 2005), few hotels are maximizing the usage of technology including hardware connectivity, storage space, and processing speed. There was a challenge of scepticism in the adoption of e-commerce by companies including tourism enterprises due to the fear of hackers' interception in this district of Melaka district in Malaysia (Ibrahim et al., 2020).

Continuous improvements in technology in tourism enable tourists to make meaningful decisions due to confidence in the security of their transactions being free from cyber-attacks when dealing with relevant tourist destinations of choice, and their digital technology interaction confidence (Stylos, 2020). Smart tourism dimensions, such as social value, circular economy, life quality, and sustainability, should also be incorporated when adopting digital innovations such as 4.0 technology in tourism, and ecotourism to enhance experiences in tourism to ultimately achieve a competitive advantage over smart tourism destinations (Pencarelli, 2020). Tourism 4.0 technologies are encouraged to be catalysts for tourism practices improvement, only when embracing smartness and

---

P. Chihwai (✉)
Vaal University of Technology, Vanderbijlpark, South Africa

E. Ndhlovu et al. (eds.), *Tourism and Hospitality for Sustainable Development*,
https://doi.org/10.1007/978-3-031-63069-9_3

sustainability as variables for enhancing life quality and the social value of tourists and local natives (Pencarelli, 2020).

Fragmented sides of digital technology in tourism have been covered by researchers and not from all-inclusive perspectives. For example, some studies have covered digital technology assisting management in hotel decision-making in Hong Kong (Law & Jogaratnam, 2005), whilst in another study customers purchase experiential tourism decision-making due to digital technology (Stylos, 2020). Meanwhile, Technology 4.0 fusion with smart tourism dimensions is enhancing tourism competitiveness covering tourism in general (Pencarelli, 2020), while another study dealt with Technology 5.0 (Pillai et al., 2021) and its effect on the hospitality industry in general.

There are limited studies, if any that specifically dealt with digital technology in the context of South Africa National Parks management, Southern Sun hotels technology, airports, and airline technology to establish how digital technology has been practised in South African tourism in one study. Given the aforementioned, this study covers the knowledge gap by establishing the level of usage of digital technology covering South African National Parks, hotels using Southern Sun Hotel group as a case study, and airports and airport aviation sectors of the tourism industry in one study, which makes this study unique, contributing to the new body of knowledge. The acknowledged knowledge gap (Kalia et al., 2022) revealed that few scientific interests and collaborations on tourism digital technology requiring further insights and inquiry reside in Africa and Russia contexts.

This chapter comprises different sections. Section 1 covered the introduction and background of the evolution and adoption of digital tourism into the tourism sector with an emphasis on the South African tourism industry. Section 2 covered related literature on tourism and digital technology synergy. Section 3 covered the materials and methods employed in the study whilst Sect. 4 dealt with findings, discussions, and conclusions.

3.2 Literature Review

Digital media is different from traditional media in many ways because the former involves what audiences download, read, and consume whilst traditional media needs new forms involving communication, digital and cultural assimilation, and prowess (McLoughlin, 2011). The new digital world demands literacy on social media, information, creative networks, social networks, information, communication, and technological skills to utilize and enhance the adoption, usage, creativity, and sharing of digital resources and capitalizing on its flexibility (McLoughlin, 2011). Simply put, digitalization enables the migration of analogue formats and processes to higher levels where the formats are readable by machines (OECD, 2020). Socially constructed meanings are made possible through the everchanging web world transitioning from Technology 2.0 (McLoughlin, 2011), through to 4.0 (Kononova et al., 2020) and 5.0 (Pillai et al., 2021) and the landscape of social networks, editing, audio, and text all enhancing sharing of knowledge, communication, and community participation.

A sustainable and responsible digital tourism ecosystem has been enabled by the internet through various abilities a tourist can execute online such as the freedom to post images of a resort area or tourism destination freely, choosing a destination from abundant choices available online, and information-seeking (Almeida-Santana et al., 2020; Camilleri, 2019). Every travel ecosystem has been enhanced with digitization such as room service, travel sales, seeking information, reservations, proposals, post-travel consumer travel, breaking boundaries in the process and redefining marketing strategies of destinations, empowering the tourists in this new economy sharing digital environment (BBC, 2020). Tourism experience has been greatly improved through the fusion of digital content into the real world (Adeola & Evans, 2019; Benyon et al., 2014). Digital Tourism Business (DTB), a new angle of viewing the digital world merging with reality in the marketing and sales platform, is receiving less attention from academics (Saura, et al., 2020) yet most researchers focus on customer engagement, branding, and market search applying digital technology in the tourism industry. Tourist experiences with technology have reduced seeking information from physical tourist facilities and the use of agents to direct interaction online (Tan et al., 2017; Wang and Wang 2018), due to the abundance of digital tools.

The dispensation of the Digital Economy program of the Russian Federation would only be made possible by incorporating digital development and digitizing, tourism included (Nyathela-Sunday et al., 2022). In tourism, it is important to consider the right terminology and definitions for digital advancement so that the associated variables for each terminology will have their subsets of identity for drivers of the tourism ecosystem and the ultimate conception and development of relevant models of management (Nyathela-Sunday et al., 2022).

Stakeholders have taken a keen interest in digital technologies that have overhauled and brought wholesome changes to the tourism sector overwhelmingly (Kalia, et al., 2022). Furthermore, whilst many publications on digital technology have taken place, it is clear there are a lot of missing links to the bigger picture following a bibliometric analysis of 120 publications. Emerging trends in academic research interests related to digital technology are skewed in favour of consumer behaviour, smart destinations management, sustainability, technology, marketing, and communication catalysts of the internet. Limited research is evident in Russia and Africa regarding digital technology.

3.3 Materials and Methods

The qualitative study was employed using the critical document review method. The main sources consulted informed the findings and implications of the study. The Annual Reports were selected because they cover digital innovation aspects that are relevant to the topic and aim of the study contain reliable information based on the

Table 3.1 Major source contributors

Year	Source
2021	South Africa National Parks Annual Report
2019	South Africa National Parks Annual Report
2022	Airports Company of South Africa Annual Report
2022	Southern Sun Annual Report
2021	Southern Sun Annual Report
2020	Chaturuka, et al.—Journal Article
2016	Guiu A.G. & Ruben, M.S.—Journal Article
2019	Mhilu, J. & Lyimo, B. J.—Journal Article
2020	Pencarelli—Journal article
2021	Pillai et al.—Journal article

Source: Author

study findings and are a springboard for the implications of the study. Databases such as Google Scholar, ResearchGate, and Web of Science were searched using keywords such as digital technology, tourism, hotel digitization, tourism digitization, airports, and air transport digitization in general. A total of 120 articles were considered with relevant topics. Further exclusion of 80 articles was based on those articles that did not address adequately the themes in the abstracts and conclusions. A further 20 articles were excluded based on those articles that did not adequately address the study themes in the findings and discussions which led to the final inclusion of 20 articles for final analysis. The next step was concentrating on such keywords in the context of South Africa. The selected articles were based on relevance to the themes of the study which are digital innovation and the tourism sector. Articles with similar corroborative findings in this study were considered more relevant to synthesize the study. More details of the references are found in the reference section.

The major sources consulted are listed in Table 3.1. The major sources are a combination of company reports such as Annual Integrated Reports from South Africa National Parks Annual Report, Airports Company of South Africa Annual Report, and Southern Sun Annual Report. Few of the peer-reviewed articles consulted from journals were utilized covering themes in the study are listed in Table 3.1.

3.4 Findings and Discussions

Digital innovation has been at the centre of discussions in tourism and hospitality for a long time. Implementation of such technological solutions has remained a huge challenge due to different reasons such as lack of funding, incompetent personnel employed, lack of innovation, and continuous improvement as an ethic of tourism enterprises leading to little progress and support from higher echelons of management. The study had interesting findings in the tourism industry regarding digital conception, adoption, and evaluation thereof.

3.4.1 Air Transport and Airports Digital Technology Strategies Employed in South Africa

The study found that the Airports Company of South Africa has massive plans for digitization at its airports by the year 2025. This proves the current inadequacy of digital innovation at major South African airports such as OR Tambo International Airport, Cape Town International Airport, King Shaka International Airport, Bram Fischer International Airport in Port Elizabeth, and Upington International in East London.

The study found that plans are afoot to do trials on smart security lanes at OR international airports. There are also plans for a self-service check-in kiosk. The Airports Company of South Africa also plans to implement self-boarding e-gates and baggage drops. To improve customer experience, measures are being planned for increased information digitization and information support systems for potential passengers and passengers seeking information. There have been debates in the airline and airport industry regarding the metaverse recently in which there has been encouragement to devise methods of increasing human interaction with digital objects at these facilities and improving digital representations of the passengers and other stakeholders in the industry. There have been debates in the airlines and airport industry regarding the metaverse recently. Virtual reality will bring a wholesome good experience exceeding the singular Oculus Virtual Reality headsets recently being tested and seen as a perfect fit example of where the industry is heading. The Airports Company of South Africa also endeavours to common-use terminal equipment to provide access to airport information from one digital interface. Central to the success of the Airports Company of South Africa (ACSA) is the Technical Services and Solutions Division (TSS). The department facilitates the conception, development, and adoption of innovative solutions to ensure experiential passenger world-class feeling is achieved at a minimal cost. The findings of this study resonate with a similar study (Guiu & Ruben, 2016) at Barcelona Airport in Spain which found that the airport passenger experience has to be improved and smart technology would be the solution at different stages such as at the checking of documents section, the baggage dropping off zone, re-booking of flights section, zone of the passengers recovering their baggage, the section of self-boarding, the ABC control zone, improvement of frequent and normal travellers' database, and the application and improvements in mobile phones applications at the airport.

The study found that the ACSA must ensure enough cyber security is in place to protect stakeholders through the advancement and sophistication of its technological systems. Several measures were put in place including the employment of qualified experts in cyber security and company-insured cyber security to avert the criminals' advanced technology. The company hatched a compliance plan to deal with the Protection of Personal Information Act (POPIA) among other regulatory instruments such as the General Data Protection Regulation (GDPR). Furthermore, the Payment Card Industry Data Security Standards (PCI DSS) were installed whilst the Minimum Information Security Standards (MISS) were also mooted to further strengthen security in this information security. The organization also put in place a cyber-security strategy and ensured its implementation. Additionally, enough tools and technologies were put in place to avoid data leakages and connectivity incidents. Passengers, with the application of virtual reality and augmented reality could visualize and experience trips that are upcoming such as what Emirates Airlines did in trying to test this technology. Emirates, a classic example of the epitome of customer experience, introduced a preview of web virtual reality on its applications which immersed a passenger into almost reality through the 3D and 360-degree view of cabin crew interior cabins. This information is made available to passengers before boarding to ensure better purchase decisions and boarding experience. To seal the security of airport operations strong and peculiar usernames and passwords authenticating the users and access authorization to the system were all put in place. The findings corroborate with those in the company's secondary data sources (ACSA Annual Report, 2022).

The study found that digital transformation took place at airports in a move described as digital transformation to avoid risks associated with such. The ACSA had to put a digital strategy in place and align with different areas of the organization. Such business areas in which the digital transformation formed a synergy which includes the knowledge and innovation strategy. The synergy had to accommodate the enterprise security strategy also. There was connectivity also between digital transformation and enterprise risk management because of the speciality of this department. There was also a synergy with the business continuity framework. In charge of this information transformation was the Information Management Committee which oversees the proper implementation of the information management plan whilst the relevant Information Technology tools, equipment, plans, and process and all architecture are in place. The findings correlate with the secondary data company sources (ACSA, 2022).

The study found that robot usage in South Africa's tourism sector has not been used yet other countries have started using them for other duties and responsibilities. Robots have been used in different airports although it is still in the infancy stage. For example, The Hong Kong International Airport also utilized technological systems to support its various stakeholders including thirteen independent cleaning robots. The robots can clean carpets, floors, operational areas at airports, and toilets sterilization. The findings corroborate with the secondary data company sources (ACSA, 2022).

The study found that Metaverse technology could bring more collaboration and benefits to various stakeholders (such as was tested on KLM Royal Dutch Airlines) within the air transport industry such as training cabin crew members, maintenance of aircraft, remote workers platforms, operations at airports and the design of airports. But the use of the metaverse could expand way beyond stimulating consumer demand, and into use cases like creating a collaboration platform for remote workers, cabin training, aircraft maintenance, design, and airport operations. An application recently launched by KLM Royal Dutch Airlines named Glue allows remote stakeholders and participants to collaborate with provisions like presentations, whiteboards, web pages, and sharing of videos, much the same as would take place in the real practical world, in this 3D virtual space. The findings also resonate with similar findings where metaverse technology has been influential in enhancing positivity, efficiency, and effectiveness within the air transport sector (Njoku et al., 2023).

The implications of the air transport industry findings in this study reveal that the new technologies would ensure more theoretical areas of research in the future by researchers in search of new and better knowledge to manage customer experiences and collaborations within the industry. Current researchers would edify digital knowledge being conceptualized and tested bringing more critical thinking and think tanks into the academic fora. The practical implications of the findings reveal that a safer airport with high-security features would be desirable and embodied by various stakeholders. Enhanced passenger experiences will be guaranteed with interactive, sophisticated yet enticing journey experiences.

3.4.2 SANParks Digital Technology Strategies in South Africa

The study found that the South Africa National Parks management for all National Parks in South Africa digitized to a lesser extent through the usage of the Management Effectiveness Tracking Tool (METT). The International Union for Conservation of Nature (IUCN) World Commission on Protected Areas defines the effectiveness of management in this context as evaluating how a protected area is being managed, using metrics such as to what degree the values of the protected are being achieved and to what degree the strategic objectives of the protected area are being articulated. METT enables the protected area management and SANParks management to manage the facilities better through informed decisions such as assessing the management progress over a given period, improving strategies of conserving protected areas, adjustments to budgets already allocated especially the expenditure side, Protected area effectiveness assessments, such as the Management Effectiveness Tracking Tool (METT) assessment enables organizations to refine conservation strategies, formulating strategies to deal with the organizational threats and weaknesses. The new technology helps management in trend analysis over a certain period. METT helps SANParks to manage its national parks effectively including cultural heritage and protected areas. SANParks, to show its commitment to this management greatness and achievements, improved the METT-SA 3 system which superseded the management target from sixty-seven per cent to seventy-one per cent whilst eight national parks achieved seventy per cent, nine of the national parks had an average of between sixty-seven and seventy percentage and Namaqua National

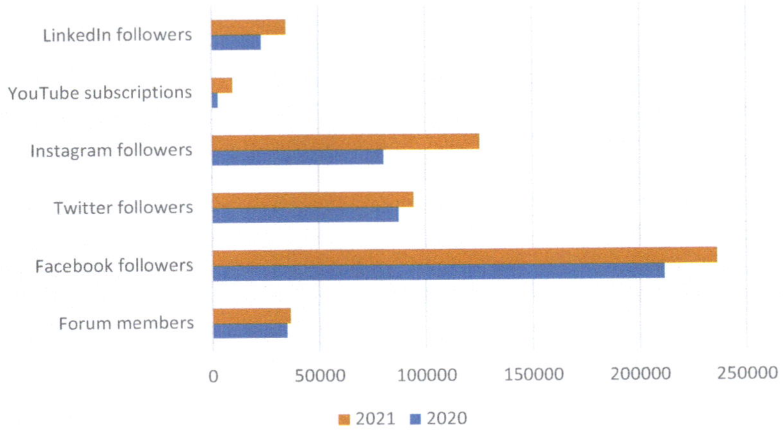

Fig. 3.1 Social Media Utilization by SANParks. (Source: Author data from SANParks Annual Report 2020, 2021)

Park achieved below sixty-seven per cent. All nineteen national parks under SANParks were assessed by the revised METT-SA 3 system Thus, during the past year, METT assessments were conducted in all parks using the revised METT-SA 3 system. During the evaluation stage and reports writing after the METT-SA 3 system suggestions for improvement were tabled and, in the reassessments, all National parks achieved over sixty-seven per cent in the financial year of 2020 and 2021. The findings are consistent with the company reports (South Africa National Parks, 2021).

Figure 3.1 shows the utilization of social media by SANParks' Marketing department from March 2020 to March 2021 which shows an increase in followers on all social media fora which also contributed to huge revenue generation.

The marketing and communication section of SANParks embraced technology in their daily operations with significant progress recently. The department utilized digital technology to improve business by reaching customers and potential travellers and increasing revenue in the process also by leveraging owned digital platforms. Innovative digital technology in the marketing department and initiatives have attracted much-needed tourists and revenue such as YouTube, Mobile technology, and other virtual reality digital social platforms creating new relationships and resuscitating the old ones. The online campaign utilized by the marketing department was 'Welcome Back to The Wild' aligning with 'Black Friday' through an online discount incentive managed to generate four million rands five hundred and fifty thousand five hundred and twenty in the year 2020. The findings are consistent with the company reports (South African National Parks Annual Report, 2019) The online 'Expanded Senior Citizen Discounts' campaign attracted slightly above seven million rand and the online campaigns brought together earned fifteen million rands nine hundred and ten thousand, one hundred and sixty-nine. Content campaigns and themed topics attracted a greater audience online such as LinkedIn from twenty-three thousand to thirty-five thousand networks, Instagram followers grew from eighty thousand followers to one hundred and twenty-five thousand followers and sixty-nine grew from three thousand to ten thousand subscribers. SANParks website increased its forums from thirty-five thousand two hundred and ninety-nine to thirty-six thousand eight hundred and eighty-three Facebook fans equally increased by 10.4 per cent, Twitter grew by 7.8 per cent, Instagram followers increased by 35.7 per cent YouTube increased by 69.3 per cent. All this shows the rigorous effort by SANParks in commitment to digital technology for the financial year of 2020 to 2021. The figures bear testimony to the importance of digital technology in enhancing business performance. Resonating with social media findings in SANParks, was a study in Tanzania national parks which showed that Facebook and Twitter are the most effective in attracting tourists to national parks (Mhilu & Lyimo, 2019). In the same study, Mhilu and Lyimo (2019) recommended future studies to assess the economic benefits of social media usage in national parks which this study found as beneficial through the accumulation of more than fifteen million rands that SANParks benefited from the online campaigns using social media.

3.4.3 *Southern Sun Hotels Digital Technology Strategies in South Africa*

The study found that there was average usage of social media in Southern Sun from the year 2020 to 2022 with a slightly more increased usage in 2022. The average monthly visits on Southernsun.com were 282,000 in the year 2022 from 217,000 in the year 2021. Figure 3.2 shows social media followers and Facebook likes for Southern Sun hotels between the years 2020 and 2022.

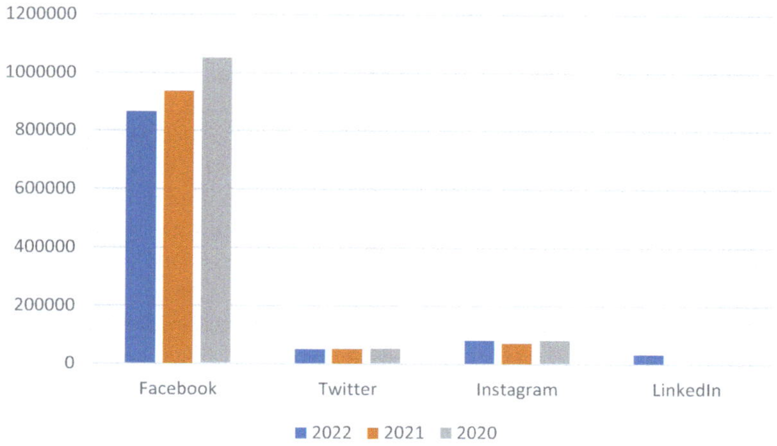

Fig. 3.2 Social media followers for Southern Sun hotels. (Source: Author data from Southern Sun Annual report)

Figure 3.2 shows that more users in the year 2020 had one million and fifty-two thousand four hundred and ninety-five followers can be attributed to the fact that Southern Sun had not yet separated its followers from the Tsogo Sun website whilst the year 2022 had 866,471 followers. Instagram was the second most used social media platform by followers with followers rising from 73,183 in the year 2021 to 82,946 in the year 2022. The least used social media platform was LinkedIn with figures not having been tracked for the consecutive years of 2020 and 2021 with a record of 34,762 in the year 2022.

The study found that the invention and adoption of digital technology in Southern Sun hotels is still at a medium level and more still needs to be done compared to the Sun International group of hotels which is more advanced. Southern Sun Hotels envisages the application of digital technology to some of its strategic business units such as reservation systems, customer reward operations, property management, procuring goods and services, and compliance in all its various forms. The findings are consistent with the company reports (Southern Sun Annual Report 2021). Furthermore, Southern Sun aspires to the utilization of technology in brand assurance, patrons' facilities, search engine optimization, total surveillance, and accommodation bookings by prospective customers including current clients and loyal customers. Continuous investment in technology is encouraged in the hotel group to ensure the conception, development, and implementation of technology in the group and to stay relevant to new technological trends in the hotel industry. The technological trends include information protection of various stakeholders such as associates, distributors, customers, and patrons, ensuring information technology business models that are cheaper are in place, ensuring robust broadband availability, and ensuring employees concerned with operations and guests have full and advanced functionalities. The advanced functionalities include secured communications and transactions meaning secrecy and confidentiality are guaranteed, ensuring equitable reward matching the value through purchased integrated tools. The findings resonate with similar studies on tourism technology adoption to enhance performance (Chihwai & Dube, 2023). The findings also resonate with similar studies on technology utilisation in hospitality and tourism (Chihwai et al., 2023).

Reservations and front office applying technology such as the PABX traditional system now being renewed to a new system remain productive and fit for purpose although upgrading enhances its relevance to POPI and GDPR requirements. The current upgrade taking place also enhances resolving operational matters with huge success. The new upgrades resulted in better business functionality, building better relationships, and customer interaction with success. The group's digital platform including but not limited to the newly re-launched southernsun.com continues to enable better customer engagement, stakeholders' relationships, and business management, especially with the new cloud-based PABX telephonic system. In meeting the sustainable development goals are the plans to introduce the cloud-based PABX in supporting the greener environment replacing the PABX electricity consumption and its footprint. Cyber security is a new trend in ensuring information security from within and without through continuous solution improvements and firewalls with double layers, in addition to appointing a partner who manages the cyber-security operations centre. This cyber-security operations centre embraces other salient sections within the company such as management vulnerability, management, detection of security threats, and how to respond to threats in all spheres of the organization. The findings are consistent with the Southern Sun Annual Report 2022. The findings resonate with similar findings on Kenya International Airport security upgrades (Nzioka & Ochieng, 2023).

3.4.4 *Synthesis of the Parks, Airports, and Hotels Through Digitization Adoption*

Figure 3.3 shows the effect of digitization on Airports, National Parks, and hotels. Digitization has the potential to reconfigure how airports can improve their efficiency and effectiveness through the introduction of technology such as having e-baggage dropping zones, self-check-in kiosks, and virtual reality experiences. The result of such initiatives will be customer and stakeholder satisfaction, improved profitability, and enhanced image and reputation. The adoption of digitization in National Parks will lead to strong management-driven technology with tangible and traceable results, improvement in social media content uploading, and management thereof which leads to stakeholder relationship management. Hotels' adoption of technology will improve the efficiency and effectiveness of their operations and management through the deployment of robots, and interactive websites which lead to customer satisfaction, enhanced image and reputation, and higher profitability.

3.5 Conclusion

Digital technology has been at the epitome of business success complementing traditional business practices, especially in tourism, which can be regarded as a highly information industry. Whilst digital technology is still in its infancy stage of usage in tourism more could be done to enhance productivity and memory for continuity. Digital technology has been used in different facets of the tourism industry to improve performance. Digital technology has been the real reason why SANParks has achieved huge strategic and non-strategic objectives. Hotels also use digital technology to achieve better performance. Airports and air transport in tourism also need to utilize information technology in all its various forms to enhance productivity and efficiency.

The COVID-19 pandemic accelerated technology fusion in the tourism industry improving convenience and reducing contactless yet good service through deploying robots, different applications on phones, artificial intelligence, and QR codes. Some of these technologies have been lying dormant but the pandemic pressed the panic button and activated awareness and the need for applications of such technologies, tools, and devices.

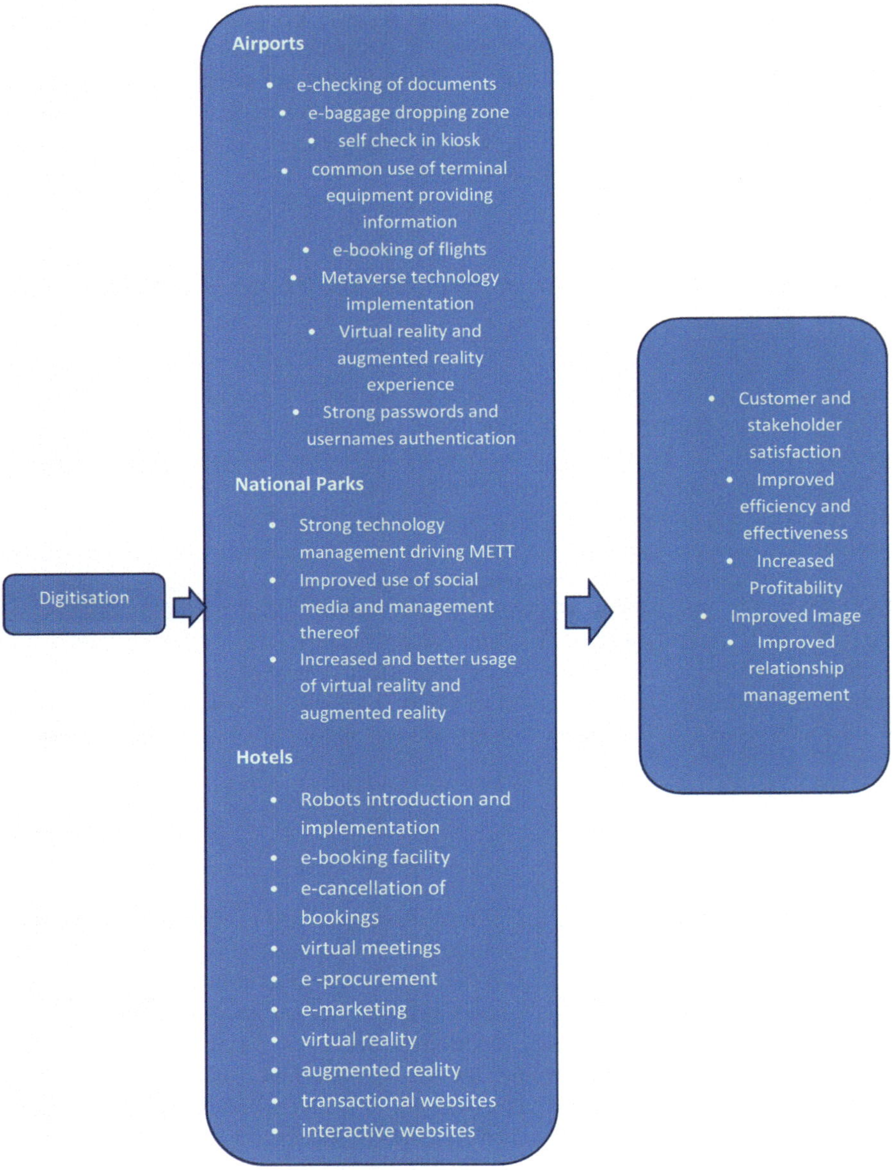

Fig. 3.3 Effect of digitization on airports, national parks, and hotels. (Source: Author)

The implication of the study entails that IT integration into all tourism systems and activities will reduce management pressure of producing high profits with few resources. Profitability achievement with little resources available can become possible if management appreciates technology and its applications not merely transferring manual work to electronic computers and employing high information technicians in the company to help upgrade and interpret the information being generated because the information is the basis for proper decision making in the tourism sector. Further implications are that closing the gap between technologically savvy young managers and the old managers and forming a good synergy will always bring better results and performance in the tourism sector which demands commitment and enthusiasm on both levels of management to steer tourism to better higher heights. Areas of future studies would be interrogating to what extent has technology been adopted in the tourism sector and establishing the challenges facing the tourism sector in the implementation of such technologies, with feasible recommendations for technology adoption.

References

Adeola, O., & Evans, O. (2019). Digital tourism: Mobile phones, internet and tourism in Africa. *Tourism Recreation Research, 44*(2), 190–202. https://doi.org/10.1080/02508281.2018.1562662

Airport Company of South Africa Annual Report. (2022). https://www.airports.co.za/Documents/ACSA%20IR%202022_Web.pdf

Almeida-Santana, A., David-Negre, T., & Moreno-Gil, S. (2020). New digital tourism ecosystem: Understanding the relationship between information sources and sharing economy platforms. *International Journal of Tourism Cities, 6*(2), 335–345. https://doi.org/10.1108/ijtc-09-2019-0173

BBC. (2020). How digitalisation is revolutionising the travel industry. http://www.bbc.com/storyworks/travel/the-new-tourismtrend/technology-redefine-tourism-industry

Benyon, D., Quigley, A., O'Keefe, B., & Riva, G. (2014). Presence and digital tourism. *AI & SOCIETY, 29*(4), 521–529. https://doi.org/10.1007/s00146-013-0493-8

Camilleri, M. A. (2019). The SMEs' technology acceptance of digital media for stakeholder engagement. *Journal of Small Business and Enterprise Development, 26*(4), 504–521.

Chaturuka, M., Duffett, R. G., & Haydam, N. (2020). Crime perceptions among international leisure tourists in Cape Town. *Safer Communities, 19*(4), 161–181. https://doi.org/10.1108/SC-04-2020-0016

Chihwai, P., & Dube, K. (2023). Exploring COVID-19 recovery mechanisms of the accommodation sector in South Africa. In K. Dube, O. L. Kupika, & D. Chikodzi (Eds.), *COVID-19 tourist destination and prospects for recovery*. Springer. https://doi.org/10.1007/978-3-031-28340-6_16

Chihwai, P., Ndlovu, E., & Dube, K. (2023). Exploring trends of South African tourist accommodation occupancy during COVID-19. In K. Dube, I. Mensah, & L. Chapungu (Eds.), *COVID-19, tourist destinations and prospects for recovery*. Springer. https://doi.org/10.1007/978-3-031-24655-5_4

Guiu A. G., & Ruben, M. S. (2016). *Feasibility study of introducing smart technologies in Barcelona Airport, Report*. GRETA,4B.

Ibrahim, I., Ismail, M. F., Amer, A., Baharuddin, F. N., Sarkun, M. R. N., & Alam, S. (2020). E-business development and the factors affecting its application in hospitality operations sustainability. In *ICIC 2020: Proceedings of the 1st international conference on Islamic Civilization, ICIC 2020, 27th August 2020, Semarang, Indonesia* (p. 22). European Alliance for Innovation.

Kalia, P., Mladenović, D., & Acevedo-Duque, Á. (2022). Decoding the trends and the emerging research directions of digital tourism in the last three decades: A bibliometric analysis. *SAGE Open, 12*(4), 21582440221128179.

Kononova, O., Prokudin, D., & Tupikina, E. (2020). From e-tourism to digital tourism. *Terminologically review*. In *SSI* (pp. 164–177).
Law, R., & Jogaratnam, G. (2005). A study of hotel information technology applications. *International Journal of Contemporary Hospitality Management, 17*(2), 170–180.
McLoughlin, C. (2011). What ICT-related skills and capabilities should be considered central to the definition of digital literacy? In *EdMedia+ Innovate Learning* (pp. 471–475). Association for the Advancement of Computing in Education (AACE).
Mhilu, J., & Lyimo, B. J. (2019). Social media marketing on attracting tourists: A case of Tanzania National Parks - Arusha. *Olva Academy – School of Researchers, 2*(3), 21.
Njoku, J. N., Nwakanma, C. I., Amaizu, G. C., & Kim, D. S. (2023). Prospects and challenges of Metaverse application in data-driven intelligent transportation systems. *IET Intelligent Transport Systems, 17*(1), 1–21.
Nyathela-Sunday, T., Septoe, N., Menze, C. Z., Banoobai-Anwar, I., Seager, D. B., & Buser, C. (2022). What innovations would enable hospitality in South Africa to rebuild? *Worldwide Hospitality and Tourism Themes, 14*(6), 565–571.
Nzioka, T. N., & Ochieng, D. O. (2023). Innovative use of emerging biometric technology in enhancing airport security at Jomo Kenyatta International Airport in Nairobi, Kenya. *Reviewed Journal of Social Science & Humanities, 4*(1), 611–638.
OECD. (2020). Preparing tourism businesses for the digital future. In *OECD tourism trends and policies* (pp. 59–85). OECD Publishing. https://www.oecd-ilibrary.org/sites/f528d444-en/index.html?itemId=/content/component/f528d444-en
Pencarelli, T. (2020). The digital revolution in the travel and tourism industry. *Information Technology & Tourism, 22*, 455–476. https://doi.org/10.1007/s40558-019-00160-3
Pillai, S. G., Haldorai, K., Seo, W. S., & Kim, W. G. (2021). COVID-19 and hospitality 5.0: Redefining hospitality operations. *International Journal of Hospitality Management, 94*, 102869.
Saura, J. R., Reyes-Menendez, A., & Palos-Sanchez, P. (2020). The digital tourism business: A systematic review of essential digital marketing strategies and trends. In J. D. Santos & Ó. L. Silva (Eds.), *Digital marketing strategies for tourism, hospitality, and airline industries* (pp. 1–22). IGI Global. https://doi.org/10.4018/978-1-5225-9783-4.ch001
South Africa National Parks Annual Report. (2019). https://www.sanparks.org/assets/docs/general/annual-report-2020.pdf
South Africa National Parks Annual Report (2020). https://www.sanparks.org/wp-content/uploads/2023/06/annualreport-2021.pdf
South Africa National Parks Annual Report. (2021). https://www.sanparks.org/assets/docs/general/annual-report-2021.pdf
Southern Sun Annual Report. (2021). https://hotels-iar.southernsun.com/reports/tsogosun-hotels-iar-2021/index.php
Southern Sun Annual Report (2022). https://hotels-iar.southernsun.com/reports/southern-sun-iar-2022/index.php
Stylos, N. (2020). Technological evolution and tourist decision-making: A perspective article. *Tourism Review, 75*(1), 273–278.
Tan, G. W. H., Lee, V. H., Lin, B., & Ooi, K. B. (2017). Mobile applications in tourism: The future of the tourism industry? *Industrial Management & Data Systems, 117*(3), 560–581. https://doi.org/10.1108/imds-12-2015-0490
Wang, B., & Wang, Z. (2018). Imported technology and CO2 emission in China: collecting evidence through bound testing and VECM approach. *Renewable and Sustainable Energy Reviews, 82*, 4204–4214.

Chapter 4
Female-Owned Fast-Food Restaurants: Technology Integration and Acceptance in Harare, Zimbabwe

Givemore Moyo, Linnet Zimusi, and Emmanuel Ndhlovu

4.1 Introduction

Technology is shaping the way business activities are being conducted across the globe. Tourism and hospitality organisations rapidly adopt and integrate novel technologies into their operations and activities (Cai et al., 2018; Triguero et al., 2013). Academics and policy institutions widely commend the rate at which tourism enterprises integrate technological innovations. For instance, Breier et al. (2021) posit that the future survival of the hospitality industry is premised upon their capacity and willingness to embrace the technological tide brought about by the advent of the Fourth Industrial Revolution (4IR). The International Civil Aviation Organization (ICAO) (2017) also argued that unless tourism enterprises rapidly embrace technological innovations, meeting the needs and expectations of key stakeholders, particularly customers, will be a significant challenge going forward. Technological innovations enable the sector to improve the quality of products, lower costs, increase efficiency, and meet customers' changing needs whilst increasing sales and profit (La et al., 2021; Ndhlovu & Dube, 2023a). Molina-Castillo et al. (2023) argue that innovation helps businesses gain market share and competitive advantage. It is also argued that the adoption of technology by restaurants facilitates their growth through increased sales, profits, market share, and competitive advantage (Ndhlovu & Dube, 2023b; Nirere, 2022; Shiwakoti et al., 2022). One of the understudied tourism and hospitality sectors is the restaurant sub-sector.

Technology plays a significant role in transforming the food services and restaurants sector in developed and developing countries (Rondović et al., 2019). The use

G. Moyo (✉) · L. Zimusi
Midlands State University, Gweru, Zimbabwe

E. Ndhlovu
Vaal University of Technology, Vanderbijlpark, South Africa

E. Ndhlovu et al. (eds.), *Tourism and Hospitality for Sustainable Development*,
https://doi.org/10.1007/978-3-031-63069-9_4

of technology in restaurants adds value to the ordinary meal and recreates experiences for customers. Customers are attracted and retained by technologies used by restaurants (Amlin et al., 2018; Berezina et al., 2019). Technology is also essential for restaurant management (Gonzalez et al., 2022). Restaurants use various technologies such as social media platforms (Luebeke, 2010), service robots (Ivanov & Webster, 2019a), point of sale (POS) systems (Koutroumanis, 2011) as well as customer-oriented self-service platforms such as tabletop multi-touch screens (Susskind & Curry, 2016).

In Zimbabwe, tourism and hospitality are two key economic sectors. According to the World Tourism and Travel Council (2023), the tourism and travel sector contributed 7.6% to the global GDP and created 22 million new jobs. This implies that the sector significantly contributes to economic empowerment, employment creation, foreign currency generation, and infrastructure development. Most small hospitality businesses in the form of lodges, restaurants, and crafts are also owned by females (Rusike & Chitambara, 2012). However, the restaurant in Zimbabwe has not attracted adequate scholarly attention. This has practical and policy implementation. In addition, the available literature on restaurants mainly focuses on developed countries, with little attention on developing countries (Ndhlovu & Dube, 2023b). Moreover, there is limited literature on the impact of technology on the growth of female-owned fast-food restaurants in developing countries such as Zimbabwe. This chapter closes this gap in the literature. This chapter (i) highlights why and how female-owned restaurants in Harare, Zimbabwe, integrate technology into their operations and (ii) explores the impact of technology on the growth of those restaurants. The chapter could interest policymakers and practitioners who could draw from the discussion and design better policies and interventions to assist restaurants in adopting better and more sophisticated technologies. Restaurant owners in Harare could also find the discussion helpful in strategising how they adopt and integrate technologies into their operations.

The chapter is organised as follows: The next section reviews the literature related to the study under various sub-sections. This is followed by the methodology adopted for the study, the presentation, and discussion of the findings. Lastly, conclusions and recommendations are drawn from the findings and the discussion.

4.2 Literature Review

The literature on technology integration in the industry has been growing in recent decades (Kindzule-Millere & Zeverte-Rivza, 2022; Kusuma et al., 2020; Marx et al., 2021). This section focuses on theoretical and empirical literature from previous studies on the technology and growth of restaurants.

4.2.1 Technology and Growth of Restaurants

Cai et al. (2018) highlight that innovation in the food industry improves the competitiveness of the players in the food industry. Carrasei et al. (2016) aver that the fast-food sector is faced with limited investment in research and development and innovation to drive performance. This view is supported by Triguero et al. (2013), who argue that the ability to grow and earn good returns results from innovation in the food sector. This shows that scholars agree that technology has the potential to assist the food sector in performing efficiently, leading to growth. For Triguero et al. (2013), technological innovation is necessary for businesses such as restaurants to survive and thrive in the contemporary, highly globalised world. Gokkaya (2020) says technology can unearth various opportunities to improve service quality, such as adequately evaluating customers and restaurant competitors. Therefore, technological innovation has come to a point whereby it is no longer avoidable by enterprises that seek to remain relevant and competitive. This is because technological innovation can reduce operational costs, decrease food waste, improve products and services, and assist in the general management of operations (Busulwa et al., 2022; Fraccastoro et al., 2021; Gokkaya, 2020).

Alawamleh et al. (2022) note that technological innovations help food companies to survive and expand in developing countries. Gebremariam and Tesfaye (2018) posit that technological innovations bring about technical changes that increase food companies' productivity and growth. Glover and Poole (2019) argue that technology has changed worldwide food processing, production, distribution, and consumption systems. Koutroumanis (2011) reveals that restaurants are realising the importance of technology in enabling them to remain competitive and, thus, are adopting novel technologies as much as possible. Chamboko-Mpotaringa and Tichaawa (2021) mention that restaurants have changed from cash registers to automated systems that help them reduce operational costs, improve products and services, and be environmentally and socially responsible. Through the integration of digital technologies, restaurants and other hospitality enterprises can reduce food waste that produces greenhouse gasses, decrease the amount of heat that they produce, and even monitor and control how they use water (Cirera et al., 2022; Tyan et al., 2021). Reducing restaurant paperwork usage has been commended as one of the first steps the sub-sector took to achieve an environmental-sensitive sector (Tussyadiah, 2020; Zeqiri et al., 2020). Koutroumanis (2011) stated that the first technology introduced by restaurants was the Point of Sale (POS) system, which enabled the automation of duties for service and kitchen staff. This technology facilitated the phasing out of handwritten orders.

Ndhlovu and Dube (2023c) also posit that due to technological improvement, customers can now place customised orders and pay for their bills from a kiosk on their table. This resulted in the provision of efficient services to customers. Yallop et al. (2023) posit that restaurants can improve their decision-making process through technology because it allows them to have information warehouses readily and utilise information and mining-related data. Hysa et al. (2021) reveal that some restaurants use social networks to improve their competitiveness. According to Ali et al. (2022), social media is a permanent technological platform that allows restaurants to communicate with their clients and increase their market presence. Fraccastoro et al. (2021), however, reveal that a lack of technological skills and knowledge in using social networks makes many enterprises shun the use of social networks.

In many tourism and hospitality enterprises, Artificial Intelligence and robots are being integrated to help improve operations. Ivanov and Webster (2019a) argue that service robots are some of the most advanced technologies ever adopted in the hospitality industry. Seyitoglu and Ivanov (2021) are more prevalent in restaurants that are located in hotels, mostly in developed countries. During COVID-19, robots became valuable tools for physical distancing between hosts and guests (Seyitoglu & Ivanov, 2021). Kindzule-Millere and Zeverte-Rivza (2022) note that robots are being created and used in the tourism industry to provide different technical capabilities and perform repetitive, dirty, and dangerous tasks as AI grows.

Tlili et al. (2021) argue that introducing service robots and related technologies has helped enterprises such as restaurants accommodate various categories of customers, including older people and persons with disabilities who can now place orders in restaurants while they are in their hotel rooms. The robot technology will deliver their order to their room. Robots can also improve the interaction between human beings and technology, which has previously been unseen in the hospitality sector (Keyser et al., 2019).

According to Lariviere et al. (2017), technology has improved the efficiency of data processing and analysis capabilities. Marinova et al. (2017) argue that technology allows hospitality company employees to understand customers better and improve job and customer satisfaction. On the other hand, Mathah and Fernando (2015) opine that whilst technology is good in the hospitality sector, it has replaced employees in sequential tasks. This implies that technology can have adverse effects in that it leads to the retrenchment of people from their jobs while it is good in that it reduces costs such as those for labour since it can perform tasks which many people perform.

The suspicion that technological innovation can result in massive unemployment is gaining momentum. According to Mhlanga et al. (2021), the eclectic concern is that the technological transition could replace the human workforce with machines, thus leading to technological unemployment. It is argued that technological transition could worsen inequality in the short term, regardless of its long-term beneficial effects (Cortis et al., 2021). Robertson (2016) asserts that the technological revolution will destroy jobs but not work. The fear of radical technological transition was also observed by the McKinsey Global Institute (2017:3), which posits that "… automation is not a new phenomenon, and fears about its transformation of the workplace and effects on employment date back centuries, even before the first Industrial Revolution in the 18th and 19th centuries."

Technology integration in the hospitality sector also positively impacts management, operations, and marketing (Ivanov & Webster, 2019b). Esposito et al. (2022) argue that new technologies are playing a significant role not only in revolutionising the dining experiences of patrons but also in reshaping how menus are designed, ordered, paid for, prepared, and delivered. The adoption of 4IR-based technologies in the restaurant sector is supported by experts who argue for the need to make significant advances in the types and quality of restaurant services to satisfy the needs of the evolving contemporary customer base (ElShafei, 2020; Frey, 2019). Technology transition is targeted towards the attraction and the satisfaction of the needs of the new generation of technophile patrons travelling on a tight budget (Esposito et al., 2022) and who significantly appreciate the personalisation of experiences and digitalisation of services.

Susskind and Curry (2016) found that customer-oriented self-service platforms such as tabletop multi-touch screens are now widely used in restaurants. They argue that these self-service platforms have led to guest empowerment. According to Fischerstrom and Akerstrom (2017), customers prefer self-service technology in fast-food and dining restaurants. Hanks et al. (2016) note that restaurants use consumers' mobile phones, tablets at the table, and touchscreens to execute various activities such as ordering food, reading the menu, customising the order, interacting with the company, and asking for refills. Brustein (2013) and Konrad (2013) report that Applebees, an American casual dining restaurant, uses tableside tablets for self-service in their restaurants to enable clients to order and pay for their meals and, after that, be served by their staff members. However, Nilsson et al. (2021) argue that there needs to be more research in restaurants concerning self-service technology, and, in the end, the customers end up using an app to order in the restaurants. However, it was found that factors such as perceived control, ease of use, usefulness, reliability, need for interaction with personnel, personal factors, technology readiness, lack of personal service, external factors, financial support, and compensation when needed affect customers' use of self-service technology (Khan, 2022; Kim et al., 2021; Lazaro, 2022).

4.2.2 Conceptual Framework

The reviewed literature indicates that technology impacts the growth of restaurants. Figure 4.1 shows the conceptual framework for the technology and growth of restaurants.

The framework shows that the adoption and integration of technologies by restaurants is the availability of Cyber-Physical Systems (CPS) (computers, laptops, tablets). In addition, such CPS should also be supported by funds needed to secure these gadgets. Restaurants will also need funds to service these gadgets from time to time. Lastly, technology adoption also depends on the Internet connection and WIFI availability. In the presence of these gadgets and services, restaurants that adopt technologies will benefit in three ways, as shown in the cells to the right of the diagram. The first is the operational growth of restaurants. This growth is in terms of improvement in revenue base, improvement of client touchpoints, market share improvement, customer retention, and improvement of customers' points of

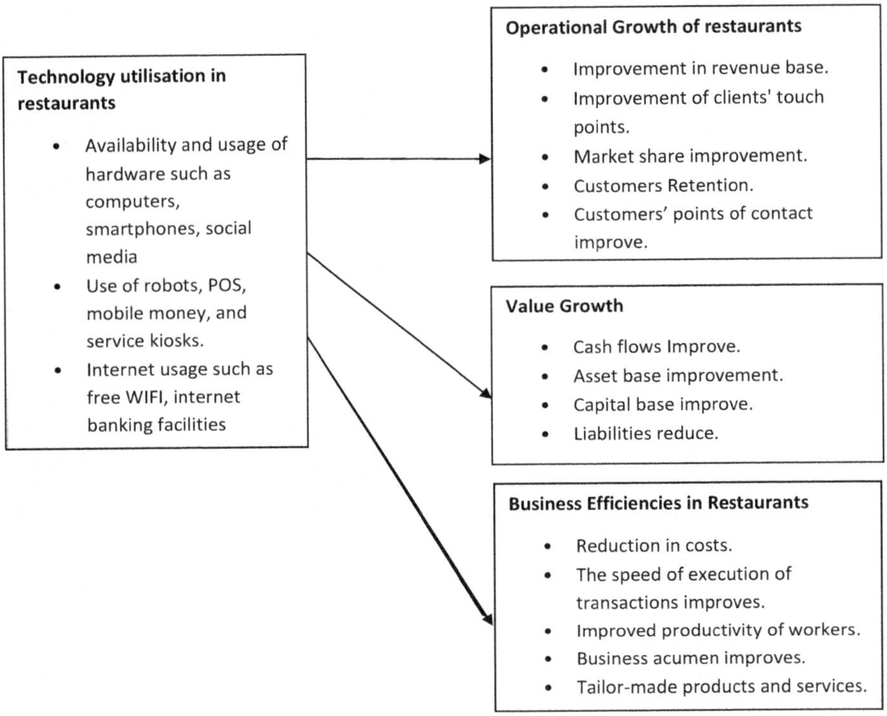

Fig. 4.1 Conceptual framework on technology and growth of restaurants. (Source: Moyo et al. (2023))

contact. Secondly, the restaurants experience value growth. This value is experienced in improved cash flows, asset base, capital base, and reduced liabilities. Thirdly, technological adoption will generate business efficiency. This will be in terms of reduced operational costs, improved speed of execution of transactions (efficiency and effectiveness), improved productivity of workers, improved business acumen, and tailor-made products and services.

Overall, the framework shows that technology adoption by restaurants impacts value growth as it determines assets, liabilities, cash flows, targeted profit meetings, and investors' attraction. It also impacts the speed of execution of transactions, the volume of transactions, employees' business acumen, and the provision of tailor-made products and services.

4.2.3 Theoretical Perspectives

This chapter deploys the Technology Acceptance Model (TAM) to interpret technological adoption and integration by female-owned restaurants in Harare. Initially proposed by Fred Davis in 1985 as a doctoral thesis at the Massachusetts Institute of Technology, current literature indicates that TAM is a highly cited model. Chuttur (2009) argues that the wide acceptance of TAM is based on the fact that the model has a sound theoretical assumption and practical effectiveness. The theoretical basis of TAM is built on the premise that when users are presented with a new technology, three significant factors influence their decision on how and when they will use it. The first is its perceived usefulness, the second is the perceived ease of use, and the third is the user attitude towards usage. According to Davis (1989), perceived usefulness is the degree to which a user believes that using a particular system would enhance his or her job performance. On the other hand, perceived ease of use is the degree to which a user believes that using a particular technology would be free from effort and not complicated. In other words, it is the degree to which users perceive technology as better than its substitutes (Jahangir & Noorjahan, 2008).

The TAM has continued to gain widespread scholarly support. For instance, Abu-Al and Steve (2013) and Glover and Poole (2019) indicate that users have different intentions to adopt and integrate technology. Intention is influenced by the perceived usefulness of technology and also the perceived ease of use. The TAM has been applied in numerous information technology and systems areas (Mugo et al., 2017). Several researchers have identified specific areas where the model has been adopted. For example, Moon and Kim (2001) used the model to explain the users' acceptance of the World Wide Web in an educational context. Chen et al. (2011) adopted the model while investigating automotive telematics users' intentions. Stern et al. (2008) used the model in their studies on consumers' acceptance of online auctions. Serenko et al. (2007) used the model to assess user acceptance of interface agents in daily work applications.

This chapter implies that the perceived usefulness and perceived ease of use affect the extent of technology usage in restaurants and the growth achieved by female-owned fast-food restaurants due to the use of technology. Many factors affect the intention of female-owned fast-food restaurants to use technology. Fast-food restaurants might be encouraged to adopt and integrate technology by its perceived usefulness to their operations. These restaurants may consider the merits and demerits of using technology before deciding to adopt them. Technology may then be adopted if it can improve the growth of restaurants. Perceived ease is also a factor that can determine the adoption of technology by female-owned fast-food restaurants. Organisations work well when they find it easier to use technology. If the technology is easier to use, restaurant owners will likely adopt and implement it. Therefore, to adopt technology, female-owned fast-food restaurants weigh the perceived usefulness and ease of use of technology.

4.3 Methods and Materials

The study adopted a sequential explanatory research design underpinned by a mixed research approach. The population of the study was all the fast-food restaurants in the Harare Central Business District (CBD). Purposive sampling was then used to target female-owned restaurants. The researchers approached each potential restaurant and, upon observing all ethical protocols required in research, sought information regarding the ownership status of the restaurants. If the restaurant was female-owned, that particular entity was recruited for participation. Restaurants that refused to be recruited were left out. The process was repeated until 30 female-owned fast-food restaurants were identified in Harare Central Business District (CBD). Data was gathered between 1 September 2023 and 15 September 2023 (2 weeks).

Targeting female-owned fast-food restaurants as participants was justified on several fundamental levels. Firstly, Zimbabwe is battling one of the worst global hyperinflations, and most businesses, including large corporations, are failing to keep pace with global competitiveness (Mhlanga & Ndhlovu, 2021). Secondly, in Africa, generally, gender norms are widely implicated by many studies as the root of many institutionalised and systemic gender inequalities that have far-reaching consequences for participation in business (Ndhlovu & Majova, 2023). Gender norms emerge from social norms—informal rules that prescribe relations between individuals and collective behaviours that shape and determine how (should) people. While gender norms are context- and cultural-specific, broad patterns and systematic ways constrain females in every community. Such constraints then directly impact how females contribute to the broader communal economy. This is also clear in Zimbabwe‘s tourism and hospitality sector, where females run most fast-food restaurants. One of the challenges is the lack of adequate resources to operate their businesses.

Within the identified 30 restaurants, the study adopted a purposive sampling technique to select 200 restaurant employees, 35 restaurant clients, and 30 restaurant managers or owners with operations. Questionnaires were distributed to a total of 200 restaurant employees. Interviews were proposed to 35 clients, of which only 25 were successful. A total of 30 interviews were also proposed to restaurant managers or owners, and twenty-five (25) interviews were successful. The interviews with restaurant owners or managers were assigned the code O/M 1-25 (for owner or manager), while interviews for customers were coded C 1-25. This was meant to simplify the data analysis process. These codes are used to identify each interview in the results presentation section.

Quantitative data was meant to be more dominant, with qualitative data only supporting it. Data analysis for quantitative data followed simple descriptive statistics, while analysis for qualitative data followed simple extraction of direct quotations that supported quantitative data. Quantitative data was summarised in the form of tables, while direct quotations and summaries were used for qualitative data.

Regarding ethical considerations, the researchers first asked permission to gather data from the owners or managers. The researchers also asked permission to gather data from the customers of female-owned fast-food restaurants. The participants were informed that data would be used for academic purposes only and would be treated with the highest degree of confidentiality. A pilot test was conducted to correct the research instruments' mistakes and ensure the validity of the questions in the research instruments. Closed questions in the questionnaires were made up of a

5-point Likert scale. SA—represented Strongly Agree, A—represented Agree, NS—represented Not Sure, D—represented Disagree, and SD—represented Strongly Disagree.

4.4 Presentation of Findings

All the twenty-five customers sampled stated that the Harare Central Business District (CBD) restaurants used the WhatsApp platform, computerised billing and receipting system, mobile money service, and point of sale services. They noted that those restaurants that used technology to display the menu and prices were larger restaurants. However, participant customers indicated that restaurants that used POS (swipe) and mobile money (Ecocash) as payment methods charged a premium, making food prices more expensive than when transacting with cash. Owners and managers also revealed that besides using digital technologies within their premises, they used them when buying stock and paying their suppliers.

Twenty-three managers stated that restaurants were forced to adopt technology due to the liquidity crisis in Zimbabwe, competition, customer demands, and the COVID-19 pandemic. One of the managers had the following words to say:

> Our customers demand improvement in the system, and you also find that other restaurants are now using technology. The Zimbabwean economic situation, where cash is scarce, has forced us to use Point of Sale (POS) and mobile money to make it easier for our clients to pay (O/M 5, September 2023).

Most employees stated that technology had affected the operational growth of restaurants by improving sales or revenue, customer retention, customers' points of contact, and market share (Table 4.1). The participant owners and managers indicated that technology integration was part of an effort to grow the business and increase the profitability of operations. This implies that these participants perceived technology as having the potential to assist their businesses in performing effectively and efficiently, thus leading to growth. This is consistent with the TAM, which espouses that some technology users will adopt technology based on its perceived usefulness. Participants' views regarding the usefulness of technology were

Table 4.1 Influence of technology on operational growth of the restaurants

	SA	A	NS	D	SD
Technology improved revenue/sales	90%	5%	0%	3%	2%
Client touch points improved as a result of technology.	60%	10%	15%	7%	8%
Customers' retention improved as a result of technology	92%	6%	0%	2%	0%
Customers point of contact improved as a result of technology	80%	15%	3%	1%	1%
Market share improved as a result of technology	91%	6%	3%	0%	0%
The business expanded as a result of technology	57%	12%	12%	10%	9%

Source: Primary data (2023)

widely backed up by the results from the quantitative survey, as summarised in Table 4.1.

Twenty-one of the female-owned fast-food restaurant managers and owners who were interviewed revealed that technology made it easier for them to transact and encouraged customers to make repeat purchases. Some participants indicated that their businesses had grown in terms of the number of clients due to digital payment platforms. One of the managers had the following words to say:

> Since we adopted technology, such as Point of Sale, where customers pay using swipe and Ecocash, the demand for our products has increased. We now have more customers, increasing our sales revenue (O/M 2, 4 September 2023).

Most of the employees who participated in this study stated that technology had affected the value growth of restaurants by improving cash flows, capital base, and the ability to attract investors (Table 4.2). This further supports the idea that users will adopt technology if they find it useful in their operations and activities.

Twenty-four managers or owners of the restaurants stated that technology had improved revenue for the restaurants and reduced costs and that this had a multiplier effect of improving the cash flows and the capital base of the restaurants. One of the managers had the following words to say:

> As a result of using technology such as POS and mobile money (Ecocash), our revenue has improved, and we have seen an improvement in cash flows and capital base. We can now expand our business and serve many customers as a result of the funds that we have generated (O/M 17, 6 September 2023).

Another participant also reported that:

> This use of WhatsApp and Facebook for advertising continues to work in our favour. Every time we have a customer, we request their phone number to notify them of specials when we have them. So, we advertise to them. Our customers also make comments on where we need to improve. This has helped us a lot to improve our sales. We are growing (O/M 19, 14 September 2023).

The employee participants reported that technology had improved business efficiencies in their respective restaurants through the reduction in operation costs, improvement in the speed of execution of transactions, the productivity of workers, the volume of transactions, the business acumen of employees, and provision of tailor-made services to the customers (Table 4.3).

One of the managers stated that:

Table 4.2 Influence of technology adoption on the value growth of restaurants

	SA	A	NS	D	SD
Technology led to the improvement in cash flows	95%	2%	3%	0%	0%
Asset base increased as a result of technology	60%	20%	2%	1%	17%
Capital base improved as a result of technology	97%	1%	0%	2%	0%
Liabilities are reduced as a result of technology	65%	10%	10%	8%	7%
The business managed to attract investors	80%	15%	2%	6%	7%

Source: Primary Data (2023)

Table 4.3 Influence of technology on business efficiencies of restaurants

	SA	A	NS	D	SD
Technology led to a reduction in costs	92%	6%	2%	0%	0%
Speed of execution of transactions improved	98%	2%	0%	0%	0%
Improved productivity of workers	75%	15%	5%	2%	3%
Volume of transactions improved	96%	3%	0%	1%	0%
Business acumen improved as a result of technology	80%	17%	6%	4%	3%
Tailor-made services were made available to clients as a result of technology	82%	16%	2%	0%	0%

Source: Primary Data (2023)

> The speed and volume of our transactions have improved. Technology is terrific because, with Ecocash or Swipe, we do not face the challenge of looking for change in an environment with a cash shortage. Our employees and I have learnt much from using technology, and our technology skills have improved. We have a WhatsApp platform where our customers pre-order meals, which has resulted in proper planning for our daily menu. There has been an improvement in the speed of serving our customers due to the automation of our systems (O/M 19, 1 September 2023).

Participants also revealed that many customers, including Ecocash and Mukuru.com, had mobile money accounts. Their customers could quickly pay using these mobile money services. One manager participant mentioned that:

> Almost everyone in this city has access to Ecocash. It is a simple service to use. They can pay with it here. We accept Ecocash. Our workers also find it very simple to use. So, we transact faster with our customers using these modern digital services (C 4, 8 September 2023).

The manager's view was also supported by several customers who indicated that they preferred using mobile transactions because they were easy to use and convenient. The quotations below summarise the views of customers:

> Mobile money is easy to use. It is just a click, and your bill is paid.
>
> Only a cellophane is what you need. You can access your account and then pay anyone. It is easy to use. No hustles at all (C11, 5 September 2023).
>
> Mobile money is faster and more efficient. You do not go through the hustle of counting bond notes to get a meal. Just a click and you go (C14, 5 September 2023).

The views of managers and customers are in line with the second factor of technology use in the TAM. The second factor indicates that users can be encouraged to use technology by its perceived ease of use. Both Managers and customers confirmed that mobile phones were easy to use. They also confirmed that mobile money services were easy to use. For this reason, they viewed digital technology as efficient and effective.

The data also showed that attitudinal issues greatly influenced users' adoption and acceptance of technology in the female-owned fast-food restaurant in Harare. Many participants accepted technology because they viewed it as safe. One owner participant revealed that

> As a woman, I find these mobile transactions safe because I do not have to keep cash with me. It is easy to be targeted. As a woman, I am vulnerable. Everyone will try to take a chance. So, I accept digital transfers to limit the amount of cash I can have on the premises (O/M 17, 5 September 2023).

Another customer also mentioned that carrying voluminous bond notes was a hustle on its own. This customer stated that:

> Bond notes are always in large packages. You will need a whole bag just to pay for a simple meal. Imagine if you will have to buy food for the family or a few friends. How much cash will you need in terms of bonds? It is too much. So, Ecocash is great for me (C9, 15 September 2023).

The attitude issue is the third factor in Davis's (1989) TAM. Thus, the attitude of female owners of restaurants also determined how they encouraged their customers and employees to engage and utilise technology.

All the managers interviewed stated that the challenges that they faced in using technology were poor network, lack of knowledge of technology by some clients and employees, and loss of personal touch between employees and customers due to technology. These challenges are partly related to the TAM's "easy to use" factor. However, they are also related to attitudinal issues, particularly with those managers or customers who view technology as detaching them from one another.

Twenty of the customers of restaurants who were interviewed also stated that the challenges they faced due to the use of technology in the restaurants included a lack of knowledge in using some technological platforms, network problems, and loss of personal interaction with restaurant staff. This agrees with Luebke (2010), who argued that lack of knowledge and challenges in using the existing social network makes many operators shun the use of social networks.

4.5 Discussion

This study found mixed reactions regarding adopting and integrating technology in female-owned fast-food restaurants in Africa. Structured according to the three factors identified by Davis (1989), the following sub-sections discuss the factors that motivate female-owned restaurants to adopt and integrate technologies in their operations.

4.5.1 Perceived Usefulness

When deploying the TAM by Davis (1989), many issues emerge regarding the reactions of customers, restaurant employees, and owners or managers. The data analysis from all categories of participants shows that the perceived usefulness of digital technologies was the primary driving force for technology acceptance and implementation in female-owned restaurants. Technologies such as mobile phones, swiping machines, and digital menus enabled restaurants to offer their customers efficient services. Digital menus provided customers with ease of access to products. The devices were avenues that provided opportunities for customers to access menus and simplified payment systems, thus increasing their enjoyment experience. This is in line with Capitanio et al., who contend that the benefits of innovation, whether in the product or process, satisfy customers' needs through quality, ease of use, security, and storage capacity. Gokkaya (2020) also adds that technology unearths various opportunities which improve service quality.

Restaurant owners and managers also indicated that technology helped them increase efficiency, reduce operation costs, and retain customers. This perceived usefulness of technology is also supported in previous studies. For instance, La et al. (2021) posit that technological innovations enable restaurants to improve the quality of products, lower costs, increase efficiency, and meet customers' changing needs whilst increasing sales and profit. Molina-Castillo et al. (2023) also aver that innovation helps businesses gain market share and competitive advantage. Ndhlovu and Dube (2023c) also argued that the adoption of technology by restaurants facilitates their growth through increased sales, profits, market share, and competitive advantage.

Technology integration by female-owned fast-food restaurants in Harare mainly adopts technologies due to their perceived potential to enable them to improve the quality of their services and products, satisfy customers, retain them, and thus remain competitive. This is how helpful technology is to them. This confirms Davis's (1989) postulation that users will adopt technology if they perceive it as applicable.

4.5.2 Perceived Ease of Use

The second aspect in Davis (1989) is perceived ease of use. In the current study, all categories of participants indicated that many of the technologies available to female-owned fast-food restaurants were easy to use. These restaurants and their customers used simple technologies such as WhatsApp platform, computerised billing and receipting system, mobile money service, and point of sale services. These were the dominant types of technologies that participants mentioned. Participants confirmed they were easy to use and explained why they did not resist using such technology. The operating systems of these technologies are designed as user-friendly systems whose operation is experiential. The user does not require formal training to operate phones or swipe machines. The methods of payment used by customers mainly involved Ecocash, a mobile money facility run by Zimbabwe's leading cell phone operator—Econet. Both customers and restaurant staff confirmed that this facility was easy to use and also served them to the huddle of carrying voluminous bond notes for purchases. Managers also commended the facility for saving them from the effort of having to count cash.

The reviewed literature also established the perceived ease of technology as one of the influencing factors for integration. Several studies show that small businesses such as independent restaurants tend to avoid complex technologies perceived as not user-friendly (Ansel & Dyer, 1999; Cirera et al., 2022; Gokkaya, 2020). In addition, customers are also likely to resist being drawn into the use of technologies, particularly if they find them hard to comprehend. However, using simple digital technologies such as mobile money transfers via Ecocash and swiping machines encouraged customers who participated in the study to accept the use of technology. This finding confirms David's (1989) observation that "perceived ease of use" was an essential determinant of whether users would accept technology.

4.5.3 Attitude Towards Use

Another issue that emerged from the findings was related to attitude towards using technologies. As indicated earlier, a user's attitude towards technology usage is essential in determining the technology's acceptance. Customers who supported the integration of technologies into the services of restaurants mentioned the issue of voluminous cash as their motivating factor. All categories of technology users in this study showed no anxiety towards technology, which often ensures when technology is not easy to use. Where technology is challenging, Gonzalez et al. (2022) observed that users often walked away in frustration, and a negative attitude was projected towards the technology. To avoid resistance, restaurant employees operated the swipe machines and only asked customers to enter their pins. By so doing, customers could change their perception towards the technology, making them embrace its utilisation.

Another issue with using technology in the participant restaurants was that their customers who used Ecocash were charged a fare. This tendency discourages customers from paying using mobile money. Customers who paid using cash, especially United States dollars, seem to receive preference ahead of those who used different payment systems. This attitude by restaurants can potentially negatively impact customers' acceptance of digital payments. On the participant customers, being charged for swiping or transferring a payment digitally raised attitudinal issues regarding digital payment systems. However, the argument can be that Zimbabwe is currently battling hyperinflation (Mhlanga & Ndhlovu, 2021). Cash is also scarce, hence the booming of mobile money businesses. Businesses that require cash, therefore, introduce an incentive for paying using cash. This could explain why customers who pay in United States dollars gain preference. The idea is not to discourage digital payments but to ensure the business has cash to run its operations. This view confirms Davis's (1989) argument that attitudes can impact users' technology adoption.

4.6 Conclusion

The chapter examined how female-owned restaurants in the Harare CBD integrated technology into their operations and explored the impact of technology on the growth of those restaurants. The chapter shows that fast-food restaurants in Harare CBD have adopted several technologies: WhatsApp, computerised billing and receipting, POS, and mobile money services. The study highlights that technology impacted the operational growth of female-owned fast-food restaurants by improving sales, customer touchpoints, and market share. The use of technology such as WhatsApp and Facebook improved product marketing. These platforms should be strengthened as part of the continued commitment to the growth of the business. POS and mobile money also provided flexible payment mechanisms for customers. In this view, these restaurants should work closely with mobile money service outlets such as Ecocash to ensure that they have updated technologies that will simplify transactions.

The adoption of technology enabled restaurants to operate efficiently and effectively. The study found that the liquidity crisis in Zimbabwe led to the adoption of technology by female-owned fast-food restaurants. Given the findings and discussion above, it is concluded that mobile money services will continue to support restaurants in overcoming many challenges, including the liquidity crisis in the country. Digital payment systems will allow clients to use various modes of payment to access products and services. Technology use will allow restaurant owners to pay for their supplies using digital technologies. Therefore, female-owned restaurants must increase technology integration to remain competitive in a volatile economic environment like Zimbabwe. Based on the conclusions above, the study recommends that female-owned restaurants teach their customers about the use of technology. This will enable the customers to use and trust the available technologies easily. There should also be regulation on the pricing when using various platforms so that customers who pay using digital platforms are not overcharged. Restaurants should also invest in advanced technologies such as digital menus and their prices to provide convenience to customers. Future studies should explore the possible funding options for small businesses such as restaurants to adopt and integrate technologies.

References

Abu-Al, A. A., & Steve, L. (2013). Factors influencing students' acceptance of m-learning: An investigation in higher education. *The International Review of Research in Open and Distributed Learning, 14*(5), 83.

Alawamleh, M., Al-Hussaini, M., & Ismail, L. B. (2022). Open innovation in the food industry: Trends and barriers: A case of the Jordanian Food Industry. *Journal of Global Entrepreneurship Research*. https://doi.org/10.1007/s40497-022-00312-6

Ali, I., Azhar, N., Tariq, M., & Khan, S. A. (2022). Role of rise in social media on travel and tourism industry. *International Journal of Research and Analytical Reviews, 9*(3), 1150–1155.

Amlin, D., Sudhahar, C., & Varghese, J. (2018). Digital technology integration for small restaurant business in India. *International Journal of Civil Engineering and Technology, 9*(6), 1593–1601.

Ansel, D., & Dyer, C. (1999). A framework for restaurant information technology. *Cornell Hotel and Restaurant Administration Quarterly, 40*(3), 74–84.

Berezina, K., Ciftci, O., & Cobanoglu, C. (2019). Robots, artificial intelligence, and service automation in restaurants. In S. Ivanov & C. Webster (Eds.), *Robots, artificial intelligence, and service automation in travel, tourism and hospitality* (pp. 185–219). Emerald.

Breier, M., Kallmuenze, A., Clauss, T., Gast, J., Kraus, S., & Tiberius, V. (2021). The role of business model innovation in the hospitality industry during the COVID-19 crisis. *International Journal of Hospitality Management, 92*, 1–12.

Brustein, J. (2013). Applebee's is now serving tablet. *Bloomberg*. https://www.bloomberg.com/news/articles/2013-12-03/applebees-is-now-serving-tablet

Busulwa, R., Pickering, M., & Mao, I. (2022). Digital transformation and hospitality management competencies: Toward an integrative framework. *International Journal of Hospitality Management, 102*(103132), 1–16.

Cai, J., Zhao, H., & Coyte, P. C. (2018). The effect of intellectual property rights protection on the international competitiveness of the pharmaceutical manufacturing industry in China. *Engineering Economics, 29*(1), 62–71.

Carraresi, L., Mamaqi, X., Albisu, L. M., & Banterle, A. (2016). Can strategic capabilities affect performance? Application of RBV to small food businesses. *Agribusiness an International Journal, 32*(3), 416–436.

Chamboko-Mpotaringa, M., & Tichaawa, T. M. (2021). Tourism digital marketing tools and views on future trends: A systematic review of literature. *African Journal of Hospitality, Tourism and Leisure, 10*(1), 712–726.

Chen, S., Li, S., & Li, C. (2011). Recent related research in technology acceptance model: A literature review. *Australian Journal of Business and Management Research, 1*(9), 124.

Chuttur, M. Y. (2009). Overview of the technology acceptance model: Origins, developments and future directions. *Working Papers on Information Systems, 9*(37), 9–37.

Cirera, X., Comin, D., & Cruz, M. (2022). *Bridging the technological divide: Technology adoption by firms in developing countries*. World Bank Group.

Cortis, N., Smyth, C., Valentine, K., Breckenridge, J., & Cullen, P. (2021). Adapting service delivery during COVID-19: Experiences of domestic violence practitioners. *The British Journal of Social Work, 51*(5), 1779–1798.

Davis, F. D. (1989). Perceived usefulness, perceived ease of use, and user acceptance of information technology. *MIS Quarterly, 13*, 319–340.

De Keyser, A., Köcher, S., Alkire, L., Verbeeck, C., & Kandampully, J. (2019). Frontline service technology infusion: Conceptual archetypes and future research directions. *Journal of Service Management, 30*(1), 156–183.

ElShafei, R. (2020). Managers' risk perception and the adoption of sustainable consumption strategies in the hospitality sector: The moderating role of stakeholder salience attributes. *Smart and Sustainable Built Environment, 11*(1), 1–18.

Esposito, B., Sessa, S. R., Sica, D., & Malandrino, O. (2022). Service innovation in the restaurant sector during COVID-19: Digital technologies to reduce customers' risk perception. *The TQM Journal, 34*(7), 134–164.

Fischerstrom, E., & Akerstrom, L. (2017). Kundrelationer i det tekniska servicemotet: Self Service Technology's betydelse for den upplevda relationen till ett foretag. *DiVA*, 1–67.

Fraccastoro, S., Gabrielsson, M., & Pullins, E. B. (2021). The integrated use of social media, digital, and traditional communication tools in the B2B sales process of international SMEs. *International Business Review, 30*(101776), 1–15.

Frey, C. B. (2019). *The technology trap: Capital, labour and power in the age of automation*. Princeton University Press.

Gebremariam, G., & Tesfaye, W. (2018). The heterogeneous effect of shocks on agricultural innovations adoption: Microeconometric evidence from rural Ethiopia. *Food Policy, 74*(April 2016), 154–161.
Glover, D., & Poole, N. (2019). Principles of innovation to build nutrition-sensitive food systems in South Asia. *Food Policy, 82*, 63–73.
Gokkaya, S. (2020). Digital transformation and marketing in tourism industry. In *Selected academic studies from Turkish Tourism Sector, chapter: 11* (pp. 149–164). Peter Lang GmbH..
Gonzalez, R., Gasco, I., & Llopis, J. (2022). Information and communication technologies in food services and restaurants: A systematic review. *International Journal of Contemporary Hospitality Management, 34*(4), 1423–1447.
Hanks, L., Line, N., & Mattila, A. (2016). The impact of self-service technology and the presence of others on cause-related marketing programs in restaurants. *Journal of Hospitality Marketing & Management, 25*(5), 547–562.
Hysa, B., Karasek, A., & Zdonek, I. (2021). Social media usage by different generations as a tool for sustainable tourism marketing in Society 5.0 idea. *Sustainability, 13, 1018*, 1–27.
International Civil Aviation Organization (ICAO). (2017). *Guidance for acceptance of Electronic Aircraft Maintenance Records (EAMR)*, 11 2017. [Online]. Available: https://www.icao.int/safety/airnavigation/OPS/airworthiness/Pages/EAMR.aspx. Accessed 27/03/2023.
Ivanov, S., & Webster, C. (2019a). *Robots, artificial intelligence, and service automation in travel, tourism and hospitality*. Emerald Publishing.
Ivanov, S., & Webster, C. (2019b). Robots in tourism: A research agenda for tourism economics. *Tourism Economics*, 1–21. https://doi.org/10.1177/1354816619879583
Jahangir, N., & Noorjahan, B. (2008). The role of perceived usefulness, perceived ease of use, security and privacy, and customer attitude to engender customer adaptation in the context of electronic banking. *African Journal of Business Management, 2*(2), 32.
Khan, M. A. (2022). Barriers constraining the growth of and potential solutions for emerging entrepreneurial SMEs. *Asia Pacific Journal of Innovation and Entrepreneurship, 16*(1), 38–50.
Kim, T. J., Almanza, B., Ma, J., Park, H., & Kline, S. F. (2021). The cleanliness of restaurants: ATP tests (reality) vs consumers' perception. *International Journal of Contemporary Hospitality Management, 33*(3), 893–911.
Kindzule-Millere, I., & Zeverte-Rivza, S. (2022). *Digital transformation in tourism: opportunities and challenges*. Proceedings of the 2022 International Conference "Economic science for rural development" No 56 Jelgava, LLU ESAF, 11–13 May 2022, pp. 476–486. https://doi.org/10.22616/ESRD.2022.56.047478
Konrad, A. (2013). Applebee's will install 100,000 intel-backed tablets next year in record rollout. Forbes. https://www.forbes.com/sites/alexkonrad/2013/12/03/applebeesintel-tablet-rollout/#7e84863c9891
Koutroumanis, D. A. (2011). Technology's effect on hotels and restaurants: Building a strategic competitive advantage. *Journal of Applied Business and Economics, 12*(1), 72–79.
Kusuma, H., Muafi, M., Aji, M. H., & Pamungkas, S. (2020). Information and communication technology adoption in small- and medium-sized enterprises: Demographic characteristics. *Journal of Asian Finance, Economics and Business, 7*(10), 969–980.
La, J., Bil, C., & Heiets, I. (2021). Impact of digital technologies on airline operations. *Transportation Research Procedia, 56*, 63–70.
Larivière, B., Bowen, D., Andreassen, T., Kunz, W., Sirianni, N., Voss, C., Wünderlich, N., & De Keyser, A. (2017). "Service encounter 2.0": An investigation into the roles of technology, employees and customers. *Journal of Business Research, 79*, 238–246.
Lazaro, M. A. (2022). Challenges of fast-food establishments Amidst pandemic. *International Journal of Academic Accounting, Finance and Management Research, 6*(3), 41–49.
Luebke, P. (2010). Social Media 101 for independent operators. *Restaurant Startup and Growth*, 17–23.

Marinova, D., de Ruyter, K., Huang, M.-H., Meuter, M., & Challagalla, G. (2017). Getting smart: Learning from technology empowered frontline interactions. *Journal of Service Research, 20*(1), 29–42.

Marx, S., Flynn, S., & Kylanen, M. (2021). Digital transformation in tourism: Modes for continuing professional development in a virtual community of practice. *Project Leadership and Society, 2*(100034), 1–11.

Mathath, A., & Fernando, Y. (2015). Robotic transformation and its business applications in food industry. In L. Zongwei (Ed.), *Robotics, automation, and control in industrial and service settings* (pp. 281–305). IGI Global.

McKinsey Global Institute. (2017). *Jobs lost, jobs gained: Workforce transitions in a time of automation*. Available at https://www.mckinsey.com/~/media/mckinsey/featured%20insights/Future%20of%20Organizations/What%20the%20future%20of%20work%20will%20mean%20for%20jobs%20skills%20and%20wages/MGI-Jobs-Lost-Jobs-Gained-Report-December-6-2017.ashx. Retrieved 21 January 2022.

Mhlanga, D., & Ndhlovu, E. (2021). Socio-economic and political challenges in Zimbabwe and the development implications for Southern Africa. *Journal of African Foreign Affairs, 8*(2), 77–100.

Mhlanga, D., Ndhlovu, E., & Hofisi, C. (2021). Assessment of the 4IR challenges of radical innovation in service delivery in Africa. *Journal of Public Administration, 56*(4.1), 1002–1017.

Molina-Castillo, F., Merono-Cerdan, A., Lopez-Nicholas, C., & Fernandez-Espinar, L. (2023). Innovation and technology in hospitality sector: Outcome and performance. *Businesses, 3*(1), 198–219.

Moon, J., & Kim, Y. (2001). Extending the TAM for a world-wide-web context. *Information & Management, 38*, 228.

Moyo, G., Zimusi, L., & Zimusi, V. (2023). Impact of information technology on the growth of Microfinance Institutions (MFIs) in Zimbabwe. *International Journal for Multidisciplinary Research, 5*(4), 1–12.

Mugo, D. G., Njagi, K., Chemwei, B., & Motanya, O. J. (2017). The Technology Acceptance Model (TAM) and its application to the utilization of mobile learning technologies. *British Journal of Mathematics & Computer Science, 20*(4), 1–8.

Ndhlovu, E., & Dube, K. (2023a). Impact of COVID-19 on tourism in South Africa: Building resilience and prospects for recovery. In K. Dube, G. Nhamo, & M. P. Swart (Eds.), *COVID-19, tourist destinations and prospects for recovery volume one: A global perspective* (pp. 81–100). Springer Nature.

Ndhlovu, E., & Dube, K. (2023b). Challenges of radical technological transition in the restaurant industry within developing countries. *African Journal of Hospitality, Tourism and Leisure, 12*(1), 156–170.

Ndhlovu, E., & Dube, K. (2023c). Restaurants industry recovery challenges from COVID-19 Amidst the Russo-Ukrainian War. In K. Dube, G. Nhamo, & M. P. Swart (Eds.), *COVID-19, tourist destinations and prospects for recovery volume one: A global perspective* (pp. 337–356). Springer Nature.

Ndhlovu, E., & Majova, B. (2023). Women empowerment in the South African agri-business: Opportunities and constraints in the Gauteng Province. In D. Mhlanga & E. Ndhlovu (Eds.), *Economic inclusion in post-independence Africa: An inclusive approach to economic inclusion*. Springer.

Nirere, B. (2022). *Challenges of digital transformation: A case study of the restaurant industry*. UMEA University. Masters Dissertation.

Nilsson, E., Pers, J., & Grubbström, L. (2021). Self-Service Technology in Casual Dining Restaurants, Services Marketing Quarterly, 42:1–2, 57–73, https://doi.org/10.1080/15332969.2021.1947085

Robertson, S. L. (2018). The role of technology in service delivery for people with disabilities in rural communities. In D. A. Harley, N. A. Ysasi, M. L. Bishop, & A. R. Fleming (Eds.), *Disability and vocational rehabilitation in rural settings* (pp. 59–76). Springer.

Robertson, N., McDonald, H., Leckie, C., & McQuilken, L. (2016). Examining customer evaluations across different self-service technologies. *Journal of Services Marketing, 30*(1), 88–102.
Rondović, B., Djuričković, T., & Kašćelan, L. (2019). Drivers of E-business diffusion in tourism: A decision tree approach. *Journal of Theoretical and Applied Electronic Commerce Research, 14*(1), 30–50.
Rusike, V., & Chitambara, P. (2012). Organising in the Zimbabwe hospitality sector. Available at http://lrs.org.za/media/2018/10/6e7e6cfd.pdf [Accessed 15 May 2020]
Serenko, A., Bontis, N., & Detlor, B. (2007). End-user adoption of animated interface agents in everyday work applications. *Behaviour & Information Technology, 26*(2), 119–132.
Seyitoglu, F., & Ivanov, S. (2021). Service robots as a tool for physical distancing in tourism. *Current Issues in Tourism, 24*(12), 1631–1634.
Shiwakoti, N., Hu, Q., Pang, M. K., Cheung, T. M., Xu, Z., & Jiang, H. (2022). Passengers' perceptions and satisfaction with digital technology adopted by airlines during COVID-19 pandemic. *Future Transportation, 2*, 988–1009.
Stern, B. B., Royne, M. B., Stafford, F. T., & Bienstock, C. C. (2008). Consumer acceptance of online auctions: An extension and revision of the TAM. *Psychology & Marketing, 25*(7), 619–636.
Susskind, A. M., & Curry, B. (2016). The influence of tabletop technology in full-service restaurants. *Cornell Hospitality Report, 16*(22), 3–9.
Tlili, A., Altinay, F., Altinay, & Zhang, Y. (2021). Envisioning the future of technology integration for accessible hospitality and tourism. *International Journal of Contemporary Hospitality Management, 33*(12), 4460–4482.
Triguero, A., Corcoles, D., & Cuerva, M. (2013). Differences in innovation between food and manufacturing firms: An analysis of persistence. *Agribusiness an International Journal, 29*(3), 273–292.
Tussyadiah, I. (2020). A review of research into automation in tourism: Launching the annals of tourism research curated collection on artificial intelligence and robotics in tourism. *Annals of Tourism Research, 81*(102883), 1–13.
Tyan, I., Yague, M. I., & Guevara-Plaza, A. (2021). Blockchain Technology's potential for sustainable tourism. In W. Worndl et al. (Eds.), *Information and communication Technologies in Tourism* (pp. 17–29). Springer.
Yallop, A., et al. (2023). The digital traveller: Implications for data ethics and data governance in tourism and hospitality. *Journal of Consumer Marketing, 40*(2), 155–170.
World Tourism and Travel Council (2023). Economic Impact Report.
Zeqiri, A., Dahmani, M., & Youssef, A. B. (2020). Digitalization of the tourism industry: What are the impacts of the new wave of technologies. *Balkan Economic Review, 2*, 63–82.

Chapter 5
Prospects of Peer-to-Peer Accommodation Platforms in Urban Tourism in Zimbabwe

Ngoni Courage Shereni and Emmanuel Ndhlovu

5.1 Introduction and Background

The recent exponential growth of Internet-based peer-to-peer (P2P) platforms has given rise to a new generation of tourism entrepreneurs (Minoia & Jokela, 2022). Small business owners, also known as micro-entrepreneurs. Are taking advantage of these platforms to bypass traditional business models by connecting customers and suppliers directly (Shabrina et al., 2021). P2P marketplaces are part of what is broadly known as the sharing economy (Giddy, 2019). This refers to the monetisation of excess capacity through short-term rental of properties like cars and apartments over digital platforms (Shereni, 2019). P2P platforms such as Uber, Airbnb and Booking.com have taken over digital tourism marketplaces, changed the way tourism actors relate and challenged the conventional way of doing tourism business (Hall et al., 2022). In the accommodation sector, these platforms enable individual homeowners to economically benefit from their properties by renting them out to travellers (Moreno-Izquierdo et al., 2019).

Globally, urban tourism is recognised as an important element of the tourism industry (Rogerson, 2011). Urban areas have always been an integral component of tourism developments leveraging on various attractions they host such as museums, art galleries, historic buildings and urban landscapes (Ndhlovu & Dube, 2023a; Shabrina et al., 2021). In the developed world, urban tourism has been used by local governments to stimulate economic development and urban regeneration (Rogerson, 2013). In this regard, Musavengane et al. (2020) note that urban tourism is attributed to the vibrancy of American and European cities and the reimaging of urban areas

N. C. Shereni (✉)
School of Tourism and Hospitality, College of Business and Economics, University of Johannesburg, Johannesburg, South Africa

E. Ndhlovu
Vaal University of Technology, Vanderbijlpark, South Africa

E. Ndhlovu et al. (eds.), *Tourism and Hospitality for Sustainable Development*,
https://doi.org/10.1007/978-3-031-63069-9_5

that have lost their attractiveness. The urban tourism landscape in the Global North is different from that of the Global South. The existence of primacy cities, a thriving informal economy as well as unclear urban, peri-urban and rural boundaries in the Global South characterise such differences (Page & Duignan, 2023). Most scholarly debates on urban tourism in Africa focus on contemporary issues around tourism development and the role of urban tourism in economic restructuring (Rogerson, 2023; Rogerson & Visser, 2014). Also, urban tourism in the African context has not yet been fully embraced as a tourism development option (Woyo, 2020).

P2P marketplaces have become drivers of change in urban areas and are leading in the development of smart cities (Minoia & Jokela, 2022). Online accommodation platforms allow temporary city visitors to stay in private apartments located in city suburbs (Stors, 2022). This has changed the dimensions of urban life and urban tourism in different ways (Stors & Baltes, 2018). Jaremen et al. (2020) observe that P2P platforms influence city functions, increase tourist traffic to cities and change the way tourism businesses in an urban environment compete. P2P accommodation platforms have given rise to what is termed *new urban tourism* where tourists leave inner city areas closer to tourist attractions in preference to staying in residential areas (Stors, 2022). This phenomenon has changed the face of residential neighbourhoods in cities both positively and negatively.

There is strong evidence in the literature on how P2P accommodation platforms have transformed urban areas in the Global North (Adamiak, 2018). However, within the Global South context, studies focusing on the influence of P2P accommodation marketplaces in urban areas are limited. Even though there is growing research of studies profiling African urban destinations in tourism spaces (Chihwai et al., 2023), the link with the platform economy is still missing. This points to the limited application of P2P marketplaces in sub-Saharan Africa. Other than South Africa most African countries do not have a visible sharing economy model that facilitates P2P marketplaces in the tourism industry (Shereni, 2019). This study, therefore, aims to analyse the prospects of P2P accommodation marketplaces in Zimbabwe. The country has a rich urban tourism resource that can anchor tourism developments. However, the use of P2P accommodation platforms in urban areas in Zimbabwe is still in its infancy.

5.2 Materials and Methods

5.2.1 Study Context

Zimbabwe is a landlocked country located in Southern Africa with a land size of 390,745 km^2. It shares borders with Mozambique in the east, South Africa in the south, Zambia in the north and Botswana in the west. According to the Zimbabwe Statistical Agency, the country had a population of just above 15 million as of 20 April 2022 (Zimstats, 2022). The national capital, Harare is the centre of government and one of the most visited cities in the country (Makoni & Rogerson, 2023; Makoni & Tichaawa, 2018). The city is considered the main gateway to the country and the centre of development in the nation (Tichaawa & Makoni, 2018: 436). Bulawayo is the second largest city located in the southwest of Zimbabwe and is regarded as the cultural hub of the country (Moyo & Tichaawa, 2017). The major attractions in the city are mainly cultural heritage dating back to the precolonial era. The two cities are important in the urban tourism developments of the country. There are however other cities in Zimbabwe such as Gweru, Mutare and Masvingo that are important resources for urban tourism.

5.2.2 *Research Design*

This study adopted a qualitative research approach based on a systematic review of literature from various sources. Google Scholar was used to search for peer-reviewed books, book chapters and journal articles that were utilised to analyse the link between P2P accommodation platforms and urban tourism in general. Search terms that were used include 'peer to peer accommodation', 'accommodation sharing economy', 'urban tourism', 'accommodation sharing economy + urban tourism' and 'peer to peer accommodation + urban tourism'. Only full-text articles written in the English language, published from 2010 to 2023, were considered for review. The search was not limited to a specific geographical context mainly because studies focusing on the prospects of P2P accommodation marketplaces in urban tourism in sub-Saharan Africa and Zimbabwe are limited. Therefore, a broader search of global literature sources would provide significant results. The search yielded a total of 4067 published articles. Duplications were identified and removed based on article characteristics such as DOI and title. The abstracts and full papers of the remaining articles were analysed to determine their relevance to this study. In the end, 323 relevant articles were retained for review.

5.3 Results and Discussion

5.3.1 *Peer-to-Peer Accommodation Marketplaces*

The emergence of the P2P accommodation marketplaces over a decade ago is considered disruptive to the conventional accommodation sector (Yeager et al., 2023). Services that were offered traditionally by professional institutionalised establishments such as hotels are now being provided by non-professional and non-institutionalised service providers through Internet-based peer-to-peer platforms (Ferreri & Sanyal, 2018). P2P accommodation platforms allow individuals to rent out second homes or individual rooms using community-based online services

(Moreno-Izquierdo et al., 2019; Ndhlovu & Dube, 2023b; Sánchez-Franco & Rey-Moreno, 2022). The platform economy has its origins in the Global North and is popular in most European cities (Moreno-Gil & Coca-Stefaniak, 2020).

Airbnb has the largest market share in the P2P accommodation business. It has been widely adopted in different geographical contexts and impacted significantly the accommodation sector globally (Hall et al., 2022). Together with other online booking platforms such as HomeAway, Vrbo, FlipKey, Couchsurfing and Roomorama, Airbnb has brought about a new ecosystem of accommodation rentals that integrates numerous actors and challenges the dominance of the brick-and-mortar accommodation establishments (Kuhzady et al., 2020). Within a destination, P2P accommodation marketplaces play an intermediation role in the commercialisation of existing accommodation places and also increase accommodation capacity at tourist destinations (Adamiak, 2020; Moreno-Izquierdo et al., 2019).

Various factors drive the adoption and use of P2P accommodation platforms. Kuhzady et al. (2020) argue that rapid technological developments, the search for authentic experiences, opportunities for socialisation, economic benefits and environmental concerns are some of the factors behind the growth of P2P accommodation. Further, personalised services provided by the host, low-cost pricing, the comfort of a home experience and opportunities for customer-to-customer interactions have a significant influence on the use of P2P accommodation (Casais et al., 2020). In the P2P platform business consumers are regarded as key players in the value co-creation process (Casais et al., 2020).

Most scholarly debates on P2P marketplaces are located in the developed Global North environments with little focus on the developing Global South regions (Giddy, 2020; Ginindza & Tichaawa, 2019; Mhlanga, 2019). Generally, Airbnb has fewer listings in Africa as compared to other countries signifying its low adoption in sub-Saharan Africa (Adamiak, 2022). The P2P accommodation platform in most sub-Saharan African countries except for South Africa and Morocco still lags in its adoption (Adamiak, 2022; Shereni, 2019). Airbnb piloted the AirBnb Africa Academy (AAA) in South Africa meant to equip the hosts with the requisite skills necessary to host guests using the P2P platform (Henama, 2021). This has resulted in the rapid growth of Airbnb business in the South African economy.

5.3.2 Urban Tourism Developments in Sub-Saharan Africa

It is estimated that 66% of the global population will be living in cities by 2050 (Brauckmann, 2017). Within the urban economy, tourism is positioned as one of the key sectors (Sans & Quaglieri, 2016). Regardless of the importance of urban tourism in the broader tourism discourse, it has received little attention from tourism scholars (Page & Duignan, 2023). Urban tourism is a complex phenomenon to understand because of the multifunctional nature of cities and also because tourism is just an extension of the economic and social activities happening within cities (Page & Duignan, 2023; Stors, 2022). Cities appeal to different travellers because of the presence of different sights, museums, universities, nightlife opportunities, trade fairs, theatres and shopping complexes among other attractions (Brauckmann, 2017). Urban tourism is therefore important to the future growth of the tourism industry.

Sub-Saharan Africa is regarded as the world's fastest urbanising region and its global share of urban residents is expected to reach 20% by 2050 (Saghir & Santoro, 2018). The rapid urbanisation in African cities presents its inhabitants with opportunities to engage in entrepreneurship, create livelihoods and accumulate assets (Awumbila, 2017). However, most scholarly debates on urban tourism have been focused on the Global North and the major focus of urban tourism studies in the Global South is South Africa and some English-speaking countries (Augusti, 2020; Musavengane et al., 2020). That said, urban tourism development patterns for sub-Saharan Africa are different from those of the Global North developed countries (Rogerson, 2011). Rogerson and Visser (2014: 408) identified five key themes emerging from discussions on urban tourism in Africa as '(1) tourism and urban economic restructuring; (2) the role of the accommodation sector; (3) slum tourism as a distinctive form of pro-poor tourism; (4) African cities as non-leisure destinations and (5) informal sector tourism'.

Township tourism has been observed to have a huge potential in countries like Zimbabwe and South Africa (Mazhande & Mamimine, 2020). Relatedly, Njaya (2014) posits that high levels of unemployment and economic challenges in urban areas have pushed residents into the informal sector, specifically street food vending, which is a common feature of urban tourist destinations in the Global South. Rogerson (2016) stressed that there is a strong link between urbanisation and the growth of the informal economy in African cities. Makoni (2017) further argue that urban tourism in the African context provides opportunities for urban dwellers, especially those in the informal economy confronted with various socio-economic challenges. Business travel and Visiting Friends and Relatives (VFR) are also predominant in African cities (Augusti, 2020). In the case of South Africa, there is a significant number of domestic tourists engaging in VFR in townships like Soweto attracted by the emergence of a strong leisure culture (Frenzel, 2020).

Urban tourism developments in Global South cities like Johannesburg are critical in spearheading economic regeneration, urban entrepreneurship and employment creation (Rogerson, 2011). However, in sub-Saharan Africa, this is a neglected niche as countries are fixated on nature-based tourism (Leornad et al., 2020; Mbaiwa et al., 2007). In the context of Botswana, poor marketing, high prices, poor infrastructure, exclusion of the residents, lack of a substantial market and lack of diversified urban tourism products are some of the challenges cited as hindering the growth of urban tourism (Mbaiwa et al., 2007). Elsewhere in South Africa, Ghana and Zimbabwe poverty, transport and infrastructure challenges as well as poor urban planning are challenges facing the viability of urban tourism (Musavengane et al., 2020).

5.3.3 Peer-to-Peer Accommodation and Urban Tourism

Recently, the resilience of urban cities in the post-COVID-19 era has become a dominant theme in urban tourism discussions (Woyo & Ukpabi, 2022). The smart cities concept is regarded as one way of constructing urban centres that are resilient to future crises (Woyo & Ukpabi, 2022). That being said, P2P marketplaces have become a critical component of smart cities (Ferreri & Sanyal, 2018; Minoia & Jokela, 2022). P2P accommodation platforms allow temporary city users such as business travellers, cultural tourists, migrants and exchange students to rent out private apartments located in residential neighbourhoods (Stors, 2022). These digital platforms have significantly shaped the way places are made, viewed and used (Minoia & Jokela, 2022). Undoubtedly, P2P accommodation platforms have changed urban life and the nature of urban tourism (Stors & Baltes, 2018).

Davidson and Infranca (2016) maintain that the growth of the accommodation-sharing economy in urban areas is anchored on finding innovative solutions to the day-to-day challenges encountered by residents. P2P marketplaces tend to thrive in urban areas mainly because of the spatial proximity and population density (Artioli, 2018). Within the context of contemporary cities, the influence of P2P accommodation platforms goes beyond tourism to impact other economic activities within the city including issues such as residents' quality of life and the administration of cities (Jaremen et al., 2020). Indeed, urban areas have become central to the development of the platform economy and they transcend private housing and short-term rental accommodation establishments (Ferreri & Sanyal, 2018). P2P platforms help to redirect traffic from tourism hotspots to areas less visited by tourists which leads urban tourist destinations to a growth trajectory (Adamiak, 2020). Sans and Quaglieri (2016) argue that among other things, platforms such as Airbnb help to revitalise peripheral neighbourhoods located away from traditional tourism areas, diversify the quality of tourism by adding the aspect of local culture to the tourist experience and help hosts to supplement household incomes.

P2P accommodation platform developments in urban areas are, however, blamed for various negative impacts such as gentrification, financialisation of housing and touristification of urban areas (Minoia & Jokela, 2022). In reality, only a handful of residents are involved in offering their houses for short-term rentals (Stors & Baltes, 2018). The majority of people participating in the P2P accommodation marketplace are non-residents of a city with the financial muscle to invest in real estate (Stors & Baltes, 2018). This triggers the departure of residents from the most lucrative and attractive areas of the city (Moreno-Izquierdo et al., 2019). In places like New York, the preference for short-term accommodation rentals is said to have resulted in a shortage of accommodation for residents and an astronomical increase in the cost of long-term rental (Jaremen et al., 2020).

In addition, the P2P accommodation platform has given rise to a parallel hospitality industry that circumvents tax and other local regulatory requirements (Ferreri & Sanyal, 2018). In this regard, various cities in the European context have enacted laws to regulate the operations of P2P platforms (Sans & Quaglieri, 2016). The loss of permanent population in residential neighbourhoods, the impact on the rental market and the inconveniences caused by the increasing leisure uses of residential buildings are some of the major reasons that prompt critics of P2P accommodation marketplaces to call for stringent regulations (Sans & Quaglieri, 2016). Repatriation of profits to corporations in Silicon Valley is blamed for revenue leakages from the national economies of developing countries (Giddy, 2019). Regardless of the identified challenges, it is envisaged that P2P accommodation marketplaces will continue to grow in importance in urban environments making platforms like Airbnb remain relevant in the future of urban tourism (Brauckmann, 2017).

5.3.4 Prospects of P2P Accommodation in Urban Tourism in Zimbabwe

P2P platforms are generally less adopted in Zimbabwe; however, the country is witnessing a proliferation of new P2P platforms of late especially in the accommodation and transport sectors. There is a huge potential for the platform economy in the Zimbabwean urban tourism space. To start with, urban areas in Zimbabwe like Bulawayo and Harare have various cultural heritage attractions that can anchor urban tourism developments (Moyo & Tichaawa, 2017). Also, colonial history embedded in townships and buildings around cities can be packaged into a viable urban tourism product (Mazhande & Mamimine, 2020). The growth of urban tourism in Zimbabwe has the potential to spur demand for P2P accommodation from tourists visiting the cities.

Nevertheless, numerous challenges encountered by the tourism industry in the country limit the potential growth of P2P accommodation. The biggest challenge for businesses in Zimbabwe is the unstable macro-economic environment which makes it difficult to run viable businesses. In the last four decades, the country has gone through different phases of economic instability (Magure, 2015). From the year 2000, Zimbabwe witnessed various episodes of hyperinflation, currency volatility and exchange rate variations. Resultantly, the economic challenges eroded disposable incomes for the locals stifling demand for tourism products (Chigora & Vutete, 2015). This poses a challenge to urban tourism development in the country and the subsequent growth of the P2P accommodation sector.

Furthermore, the country has a history of political upheaval that negatively affects tourism development (Makoni & Tichaawa, 2018). The chaotic land reform from the year 2000, Operation Restore Order/*Operation Murambatsvina* in 2005 and the violent elections in 2008 which culminated in a Government of National Unity (GNU) are notable events that led to the demise of the Zimbabwean tourism industry (Chibaya, 2013; Rogerson, 2016). In addition to that, a history of disputed elections in Zimbabwe has continued unabated since the year 2000 and the coup in November 2017 are events that caused uncertainty in the tourism business operating environment (Musavengane, 2020). Such an unpredictable political environment scares away investors, especially established companies in the P2P accommodation business. Etambakonga and Mugangu (2020) caution that tourism thrives in a stable political environment. Therefore, the aforementioned political events in Zimbabwe create a bad image for the country's tourism offering diverting potential tourists to other competitive destinations.

Sanctions imposed on Zimbabwe by Western countries in response to the fast-track land reform programme and alleged human rights abuse by the government have several impacts on tourism businesses (Ndakaripa, 2021). Firstly, they are blamed for the collapse of the economy which presents a challenging business operating environment (Chakawa, 2023). Secondly, the United States of America inhibits American companies from doing business with specified individuals and companies from Zimbabwe (Ndakaripa, 2021). This means that American companies are restricted from operating in Zimbabwe. In addition, financial transactions between Zimbabwe and other countries have to go through third parties as a result of the sanctions. This put a strain on the ease of doing business, especially for foreign companies operating in the country.

Zimbabwe has a significant informal economy concentrated in the urban areas (Rogerson, 2016), this can benefit from the P2P accommodation by providing urban residents with an opportunity to monetise their assets. Urban entrepreneurship is a common phenomenon in urban areas in Zimbabwe due to the high rate of unemployment. A number of entrepreneurs are already taking advantage of opportunities provided by platforms such as Airbnb. This boosts the adoption of P2P accommodation platforms in the urban tourism environment in Zimbabwe.

5.4 Conclusion

This chapter examines the prospects of P2P accommodation in urban tourism in Zimbabwe. There appears to be a dearth of literature on the link between P2P accommodation marketplaces and urban tourism in sub-Saharan Africa, specifically in Zimbabwe. Nature-based tourism is dominant in the tourism industry in Zimbabwe and urban tourism is not widespread. In addition, P2P marketplaces are less visible in Zimbabwe's urban settings. The prospects of P2P accommodation are however impacted by challenges that include economic volatility, political challenges and the general socio-economic environment. However, there is potential for growth of P2P accommodation within the urban tourism context of Zimbabwe. The rich cultural heritage and numerous attractions in urban areas present a viable urban tourism product that needs to be properly packaged and marketed. Locally owned P2P accommodation companies have room to occupy and dominate the market often neglected by renowned P2P accommodation companies. With domestic tourism being promoted as a viable option to international tourism which is regarded as less resilient to external shocks, the P2P accommodation business stands to benefit from developments in this market, especially within urban settings. That being said, it can be concluded that regardless of the challenges faced in urban tourism in Zimbabwe, the prospect for growth of P2P accommodation in cities in Zimbabwe is high.

This study is limited in that it is based on document analysis. The conclusions from this study are drawn from a review of literature, of which most of these studies emanate outside the context of Zimbabwe. Future studies should however carry out empirical surveys to determine the prospects of P2P accommodation platforms in developing countries like Zimbabwe.

References

Adamiak, C. (2018). Mapping Airbnb supply in European cities. *Annals of Tourism Research, 7*(1), 67–71. https://doi.org/10.1016/j.annals.2018.02.008

Adamiak, C. (2020). Peer-to-peer accommodation in destination life cycle: the case of Nordic countries. *Scandinavian Journal of Hospitality and Tourism, 20*(3), 212–226. https://doi.org/10.1080/15022250.2020.1775116

Adamiak, C. (2022). Current state and development of Airbnb accommodation offer in 167 countries. *Current Issues in Tourism, 25*(19), 3131–3149. https://doi.org/10.1080/13683500.2019.1696758

Artioli, F. (2018). *Digital platforms and cities: a literature review for urban research* (pp. 524–530). HAL.

Augusti, D. P. I. (2020). Mapping tourist hot spots in African cities based on Instagram images. *International Journal of Tourism Research, 22*(5), 617–626.

Awumbila, M. (2017). Drivers of migration and urbanization in Africa: Key trends and issues. In *Background paper prepared for UN expert group meeting on sustainable cities, human mobility and international migration, 7–8, September 2017* (Issue September). http://www.un.org/en/development/desa/population/events/pdf/expert/27/papers/III/paper-Awunbila-final.pdf

Brauckmann, S. (2017). City tourism and the sharing economy – Potential effects of online peer-to-peer marketplaces on urban property markets. *Journal of Tourism Futures, 3*(2), 114–126. https://doi.org/10.1108/JTF-05-2017-0027

Casais, B., Fernandes, J., & Sarmento, M. (2020). Tourism innovation through relationship marketing and value co-creation: A study on peer-to-peer online platforms for sharing accommodation. *Journal of Hospitality and Tourism Management, 42*(October 2019), 51–57. https://doi.org/10.1016/j.jhtm.2019.11.010

Chakawa, J. (2023). Why sanctions have not worked: Zimbabwe's experience from 2001-2021. *Third World Thematics: A TWQ Journal, 8*(4–6), 205–218. https://doi.org/10.1080/23802014.2022.2080248

Chibaya, T. (2013). From "Zimbabwe Africa's paradise to Zimbabwe a World of wonders": Benefits and challenges of rebranding Zimbabwe as a tourist destination. *Developing Country Studies, 3*(5), 84–92. http://www.iiste.org/

Chigora, F., & Vutete, C. (2015). Elasticity of Demand a Critical Form Tourism Market Position: Understanding Zimbabwe Tourism Destination. *Case Studies in Business and Management, 2*(2), 18. https://doi.org/10.5296/csbm.v2i2.8570

Chihwai, P., Ndhlovu, E., & Dube, K. (2023). Exploring trends of South African tourist accommodation occupancy during COVID-19. In K. Dube, I. Mensah, & L. Chapungu (Eds.), *COVID-19, tourist destinations and prospects for recovery*. Springer.

Davidson, N. M., & Infranca, J. (2016). The sharing economy as an urban phenomenon. *Yale Law & Policy Review, 34*(2), 215–279.

Etambakonga, C. L., & Mugangu, D. T. (2020). Urban risk and tourism in Africa: An overview. In L. Leonard, R. Musavengane, & P. Siakwah (Eds.), *Sustainable urban tourism in Sub-Saharan Africa: Risk and resilience* (pp. 1–13). Routledge.

Ferreri, M., & Sanyal, R. (2018). Platform economies and urban planning: Airbnb and regulated deregulation in London. *Urban Studies, 55*(15), 3353–3368. https://doi.org/10.1177/0042098017751982

Frenzel, F. (2020). Touring poverty in townships, inner-city, and Rural South Africa. In J. M. Rogerson & G. Visser (Eds.), *New directions in South African tourism geographies, geographies of tourism and global change* (pp. 167–181). Springer. https://doi.org/10.1007/978-3-030-29377-2

Giddy, J. K. (2019). The influence of e-hailing apps on urban mobilities in South Africa. *African Geographical Review, 38*(3), 227–239. https://doi.org/10.1080/19376812.2019.1589732

Giddy, J. K. (2020). Travel patterns of uber users in South Africa. In J. M. Rogerson & G. Visser (Eds.), *New directions in South African tourism geographies* (pp. 113–128). Springer.

Ginindza, S., & Tichaawa, T. M. (2019). The impact of sharing accommodation on the hotel occupancy rate in the kingdom of Swaziland. *Current Issues in Tourism, 22*(16), 1975–1991. https://doi.org/10.1080/13683500.2017.1408061

Hall, C. M., Prayag, G., Safonov, A., Coles, T., Gössling, S., & Naderi Koupaei, S. (2022). Airbnb and the sharing economy. *Current Issues in Tourism, 25*(19), 3057–3067. https://doi.org/10.1080/13683500.2022.2122418

Henama, U. S. (2021). eCommerce within the tourism industry in the Global South: The case of the sharing economy in South Africa. In F. Martines & D. Lopez Lopez (Eds.), *Advances in digital marketing and eCommerce* (pp. 280–290). Springer.

Jaremen, D. E., Nawrocka, E., & Żemła, M. (2020). Externalities of development of the sharing economy in tourism cities. *International Journal of Tourism Cities, 6*(1), 138–157. https://doi.org/10.1108/IJTC-05-2019-0068

Kuhzady, S., Seyfi, S., & Béal, L. (2020). Peer-to-peer (P2P) accommodation in the sharing economy: A review. *Current Issues in Tourism*, 1–16. https://doi.org/10.1080/13683500.2020.1786505

Leornad, L., Musavengane, R., & Siakwah, P. (2020). Urban risk and tourism in Africa: An overview. In L. Leonard, R. Musavengane, & P. Siakwah (Eds.), *Sustainable urban tourism in Sub-Saharan Africa: Risk and resilience* (pp. 1–13). Routledge.

Logistic Makoni. (2017). *Stakeholder reflections on Zimbabwe's Urban Tourism*. University of Johannesburg.

Magure, B. (2015). Interpreting urban informality in Chegutu, Zimbabwe. *Journal of Asian and African Studies, 50*(6), 650–666. https://doi.org/10.1177/0021909614535568

Makoni, L., & Rogerson, C. M. (2023). Business tourism in an African city: Evidence from Harare, Zimbabwe. *Studia Periegetica 43*(3), 25–48. https://doi.org/10.58683/sp.596

Makoni, L., & Tichaawa, T. M. (2018). Stakeholders' assessments of tourism policy and product development in Zimbabwe. *African Journal of Hospitality, Tourism and Leisure, 7*(4), 1–19.

Mazhande, P., & Mamimine, P. W. (2020). Developing a model for packaging and popularizing township tourism in Zimbabwe. *Hospitality & Tourism Review, 1*(2), 50–65. https://doi.org/10.38157/hospitality-tourism-review.v1i2.217

Mbaiwa, J. E., Toteng, E. N., & Moswete, N. (2007). Problems and prospects for the development of urban tourism in Gaborone and Maun, Botswana. *Development Southern Africa, 24*(5), 725–740. https://doi.org/10.1080/03768350701650793

Mhlanga, O. (2019). Peer-to-peer-travel: Is Airbnb a friend or foe to hotels? *International Journal of Culture, Tourism, and Hospitality Research, 13*(4), 443–457. https://doi.org/10.1108/IJCTHR-05-2019-0087

Minoia, P., & Jokela, S. (2022). Platform-mediated tourism: Social justice and urban governance before and during COVID-19. *Journal of Sustainable Tourism, 30*(5), 951–965. https://doi.org/10.1080/09669582.2021.1922426

Moreno-Gil, S., & Coca-Stefaniak, J. A. (2020). Overtourism and the sharing economy – Tourism cities at a crossroads. *International Journal of Tourism Cities, 6*(1), 1–7.

Moreno-Izquierdo, L., Ramón-Rodríguez, A. B., Such-Devesa, M. J., & Perles-Ribes, J. F. (2019). Tourist environment and online reputation as a generator of added value in the sharing economy: The case of Airbnb in urban and sun and beach holiday destinations. *Journal of Destination Marketing and Management, 11*, 53–66. https://doi.org/10.1016/j.jdmm.2018.11.004

Moyo, S., & Tichaawa, T. M. (2017). Community involvement and participation in Tourism development: A Zimbabwe study. *African Journal of Hospitality, Tourism and Leisure, 6*(1), 1–211.

Musavengane, R. (2020). Urban risk and tourism in Africa: An overview. In L. Leonard, R. Musavengane, & P. Siakwah (Eds.), *Sustainable urban tourism in Sub-Saharan Africa: Risk and Resilience* (pp. 1–13). Routledge.

Musavengane, R., Siakwah, P., & Leonard, L. (2020). The nexus between tourism and urban risk: Towards inclusive, safe, resilient and sustainable outdoor tourism in African cities. *Journal of Outdoor Recreation and Tourism, 29*(August 2019), 100254. https://doi.org/10.1016/j.jort.2019.100254

Ndakaripa, M. (2021). Zimbabwe's economic meltdown: Are sanctions really to blame? *The Washington Quarterly, 44*(2), 95–120. https://doi.org/10.1080/0163660X.2021.1934997

Ndhlovu, E., & Dube, K. (2023a). Impact of COVID-19 on tourism in South Africa: Building resilience and prospects for recovery. In K. Dube, G. Nhamo, & M. P. Swart (Eds.), *COVID-19, tourist destinations and prospects for recovery volume one: A global perspective* (pp. 81–100). Springer Nature.

Ndhlovu, E., & Dube, K. (2023b). Restaurants industry recovery challenges from COVID-19 Amidst the Russo-Ukrainian War. In K. Dube, G. Nhamo, & M. P. Swart (Eds.), *COVID-19, tourist destinations and prospects for recovery volume one: A global perspective* (pp. 337–356). Springer Nature.

Njaya, T. (2014). Operations of street food vendors and their impact on sustainable urban life in high density Suburbs of Harare, in Zimbabwe. *Asian Journal of Economic Modelling, 2*(1), 18–31. https://doi.org/10.18488/journal.8.2014.21.18.31

Page, S. J., & Duignan, M. (2023). Progress in Tourism Management: Is urban tourism a paradoxical research domain? Progress since 2011 and prospects for the future. *Tourism Management, 98*, 104737. https://doi.org/10.1016/j.tourman.2023.104737

Rogerson, C. M. (2011). Urban tourism and regional tourists: Shopping in Johannesburg, South Africa. *Tijdschrift Voor Economische En Sociale Geografie, 102*(3), 316–330. https://doi.org/10.1111/j.1467-9663.2011.00666.x

Rogerson, C. M. (2013). Urban tourism, economic regeneration and inclusion: Evidence from South Africa. *Local Economy, 28*(2), 188–202. https://doi.org/10.1177/0269094212463789

Rogerson, C. M. (2016). Responding to informality in Urban Africa: Street trading in Harare, Zimbabwe. *Urban Forum, 27*(2), 229–251. https://doi.org/10.1007/s12132-016-9273-0

Rogerson, C. M. (2023). Historical urban tourism: The evolution of tourism in Colonial Lourenço Marques, Mozambique c. 1920-1975. *Journal of Tourism & Development, 42*, 9–25. https://doi.org/10.34624/rtd.v42i0.32655

Rogerson, C. M., & Visser, G. (2014). A decade of progress in African urban tourism scholarship. *Urban Forum, 25*(4), 407–417. https://doi.org/10.1007/s12132-014-9238-0

Saghir, J., & Santoro, J. (2018). *Urbanization in Sub-Saharan Africa- meeting challenges by bridging stakeholders*. Center for Strategic & International Studies. http://thegreentimes.co.za/wp-content/uploads/2019/03/Urbanization-in-Sub-Saharan-Africa.pdf%0Ahttps://www.csis.org/analysis/urbanization-sub-saharan-africa

Sánchez-Franco, M. J., & Rey-Moreno, M. (2022). Do travelers' reviews depend on the destination? An analysis in coastal and urban peer-to-peer lodgings. *Psychology and Marketing, 39*(2), 441–459. https://doi.org/10.1002/mar.21608

Sans, A. A., & Quaglieri, A. (2016). Unravelling Airbnb: Urban perspectives from Barcelona. In R. A. Paolo & G. Richard (Eds.), *Reinventing the local in tourism* (pp. 209–228). Channel View Publications.

Shabrina, Z., Buyuklieva, B., & Ng, M. K. M. (2021). Short-term rental platform in the urban tourism context: A Geographically Weighted Regression (GWR) and a Multiscale GWR (MGWR) approaches. *Geographical Analysis, 53*(4), 686–707. https://doi.org/10.1111/gean.12259

Shereni, N. C. (2019). The tourism sharing economy and sustainability in developing countries: Contribution to SDGs in the hospitality sector. *African Journal of Hospitality, Tourism and Leisure, 8*(5), 1–10.

Stors, N. (2022). Constructing new urban tourism space through Airbnb. *Tourism Geographies, 24*(4–5), 692–715. https://doi.org/10.1080/14616688.2020.1750683

Stors, N., & Baltes, S. (2018). Constructing urban tourism space digitally: A study of Airbnb listings in two Berlin neighborhoods. *Proceedings of the ACM on Human-Computer Interaction, 2*(CSCW). https://doi.org/10.1145/3274435

Tichaawa, T. M., & Makoni, L. (2018). Sociodemographic influences on residents' perception of tourism development in Zimbabwe. *GeoJournal of Tourism and Geosites, 22*(2), 426–446.

Woyo, E. (2020). Urban risk and tourism in Africa: An overview. In L. Leonard, R. Musavengane, & P. Siakwah (Eds.), *Sustainable urban tourism in Sub-Saharan Africa: Risk and resilience* (pp. 1–13). Routledge.

Woyo, E., & Ukpabi, D. (2022). Building resilient smart cities for sustainable urban tourism in Africa post-COVID-19 pandemic. In *Information and communication technologies in tourism, January 11-14 2022* (pp. 256–270). https://doi.org/10.1007/978-3-030-94751-4_20

Yeager, E., Boley, B. B., & Goetcheus, C. (2023). Conceptualizing peer-to-peer accommodations as disruptions in the urban tourism system. *Journal of Sustainable Tourism, 31*(2), 504–519. https://doi.org/10.1080/09669582.2020.1864387

Zimstats. (2022). "Population and Housing Census." Zimbabwe 2022 Population and Housing Census Report Volume 1. Retrieved August 7, 2023 (https://www.zimstat.co.zw/wpcontent/uploads/publications/Population/population/phc2022/2022_PHC_Report_27012023_Final.pdf)

Chapter 6
The State of Hi-Tech Innovation in the Global Aviation Industry

Emmanuel Ndhlovu, Tawanda Makuyana, and Kaitano Dube

6.1 Introduction

The airline industry is prone to internal and external disruptions that expose its highly fragile nature (Olaganathan, 2021). The fragility requires concerted efforts to make the industry more efficient and resilient in the face of setbacks. Pandemics such as the coronavirus of 2019 (COVID-19), terrorist attacks such as the 9/11 in 2001 in the United States, the global financial crisis in 2007/8, and global energy crises have often forced or encouraged innovativeness and adoption of new business strategies by the airline industry across the world (Tan & Masood, 2021). For many airlines, new business strategies have included massive hi-tech innovation, thus, fast-tracking transformation and innovation to digitalisation (Shiwakoti et al., 2022). However, although sluggish, the transformation efforts in the industry have been ongoing for decades as airlines worked on improving efficiency and customer contentment and, consequently, their competitiveness (Kuisma, 2018). In recent years, one way to respond to disruptions by industry has been to digitalise services and integrate novel technologies. Thus, since the COVID-19 troubles in late December 2019, digital and technology-based services have gained centre stage both in institutions and intellectual cycles (Afaq et al., 2021).

Nonetheless, the aviation industry's capacity and resilience to deal with shocks remains low (Dube et al., 2021). Several studies on the issue mainly focus on the benefits and challenges of digitalisation to airlines and airports but do not explore the current state of affairs in terms of digitalisation by these enterprises. This study fills this gap which has practical implications.

Existing studies on digitalisation in the airline industry are mainly theoretical-based and conceptual, with very little empirical data support. These studies mostly

E. Ndhlovu (✉) · T. Makuyana · K. Dube
Vaal University of Technology, Vanderbijlpark, South Africa
e-mail: kaitanod@vut.ac.za

E. Ndhlovu et al. (eds.), *Tourism and Hospitality for Sustainable Development*,
https://doi.org/10.1007/978-3-031-63069-9_6

acknowledge the diagnostic potential of Artificial Intelligence (AI) and the Internet of Things (IoT) in enhancing and improving airline service quality and competitiveness (Amankwah-Amoah, 2021; Suk & Kim, 2021). Some studies focus on customer satisfaction (Moon et al., 2021; Noviantoro & Huang, 2021) and customer perceptions of digitalisation (Tyagi & Lodewijks, 2022), while others examine the role of COVID-19 in speeding up technological transformation in the industry (Olaganathan, 2021; Pereira, 2021). Thus, studies which explore the state of hi-tech innovation in the aviation industry remain scant. This article closes this gap which can have both policy and scholarly implications. The article describes the state of hi-tech innovation in the aviation industry at a global level in recent years with a focus on purposefully ten selected airlines. The study emphasises the reasons for digitalisation and technology integration. The study has the potential to conscientise some aviation enterprises on the general value of hi-tech innovation in the industry. In addition, the study could inspire further research by scholars who could confirm or problematise the descriptions presented in this article.

The article is structured as follows: The next section reviews the literature on digital adoption and technology integration before and following the COVID-19 pandemic. This section is followed by an outline of the research methodology for the study and the results and discussions. Lastly, the conclusions and recommendations are drawn from the results and the discussion.

6.2 Literature Review

The literature on hi-tech innovation in the aviation industry became more pronounced, particularly following the COVID-19 global pandemic. Therefore, this article reviews the literature on hi-tech innovation before and after COVID-19.

6.2.1 Hi-Tech Innovation Before COVID-19

The advent of COVID-19 fast-tracked the trajectory of digital transformation and technology integration by businesses worldwide. This can be demonstrated by the sudden widespread and greater utilisation of Fourth Industrial Revolution-based (4IR) 'new' technologies across industries, including aviation. However, digital transformation and technology integration trends were already underway before the arrival of COVID-19, with a broad interest in the IoT already on the rise (Apăvăloaie, 2014; Chihwai et al., 2023; Ndhlovu & Dube, 2023a, b; Shiwakoti et al., 2022). According to Cohen-Almagor (2011), the early Internet was first developed and applied in American universities, a few research units, and telecommunication firms. In the 1960s and 1970s, the programme grew to become a system of computers that conveyed data by "packet switching" (Cohen-Almagor (2011: 46). In the 1990s, the IoT expanded worldwide at an extraordinary pace and inspired impressive technological advances. Before COVID-19, companies had already integrated IoT analytics to transform big data into profits. For instance, Manyika et al. (2015) found that mining businesses already used IoT analytics to monitor their production practices, while IoT devices were installed on company fleets to track and monitor efficiency. Shiwakoti et al. (2022) also argue that edge-computing technology was common before COVID-19.

The deployment of 5G was also already unfolding before COVID-19 though adoption was low (West, 2016). Some cloud enterprises have also offered blockchain as a security facility (Haas, 2018). Deepika et al. (2020) observed that businesses were already integrating AI into their structures to reorganise processes and procedures for their staff and customers. Rotsidis et al. (2019) found that even before COVID-19, companies introduced chatbots to expedite customer support, and businesses had already learnt that Augmented Reality (AR) could advance their technological pursuits.

6.2.2 Post-COVID-19 Hi-Tech Innovation in the Aviation Industry

With digitalisation globally, the aviation industry is inexorably evolving towards increased hi-tech innovation. The reviewed literature indicates that digital technologies are being extensively implemented in different areas of the airline industry (Afaq et al., 2021; Moon et al., 2021; Noviantoro & Huang, 2021). Molchanova (2020) found that the aviation industry was rapidly integrating blockchain technology, AI, AR and VR, beacon technology, big data and analytics, and biometrics in managing air traffic, aircraft maintenance, and monitoring flights and airport control. Serrano and Kazda (2020) found that airlines had adopted digital technologies to add value to business operations. Shiwakoti et al. (2022) found that airlines' digital technologies now perform many of the tasks, such as check-ins.

Chakraborty et al. (2021) proposed a framework that highlights the rapport between the use of AI and IoT with passenger satisfaction. The framework shows that AI and IoT facilitate observing sustainable operations and improve customer confidence and satisfaction. Chakraborty et al. (2021) note that AI is being used by airlines to serve customers through official websites and social media handles through which they can timely respond to passengers' queries. IoT in airlines is integrated into the processing of E-luggage tags, E-menu, and Self-check-in kiosks which enable the collection and connection of customer data with stakeholders seamlessly in real-time (La et al., 2021; Shiwakoti et al., 2022).

Delgado-Mohatar et al. (2020) found that blockchain technology had been adopted to convert customer processing operations from physical to digital identity checks using a decentralised database. This has helped airlines to develop a vigorous security scheme by which they manage customer data and enhance biometrics recognition (El Fkharany, 2022). Delta (2021) reports that Spirit Airlines, JetBlue, and Delta Airways have already used this technology to enable customers to use facial recognition for luggage drop, security, and boarding. In Asia, AirAsia integrated AI into the mobile application—Scan2Fly, where, instead of staff, AI was used to verify the Polymerase Chain Reaction test certificate and other documents related to COVID-19 (Airasia, 2022). In Singapore, Olaganathan (2021) reports that Singapore Airlines adopted the digital Travel Pass Framework based on a digital health verification process that enables customers to store and present their COVID-19 tests and vaccine status digitally. This not only improved customer experience but also strengthened the collaboration among airlines.

In response to COVID-19 safety protocols, airlines also adopted digital contactless services. The robots spray antimicrobial agents on the cabin surface and, thus, promote observing safety protocols. In 2020, several airlines, including Honeywell, adopted an Ultraviolet Cabin to reduce the spread of viruses, such as COVID-19 and bacteria on targeted surfaces (Shiwakoti et al., 2022). Similar technologies were also implemented within the aviation industry to sanitise airports and cabins.

Hi-tech innovation was also found to increase competitiveness and profitability (Kuisma, 2018; Ndhlovu & Dube, 2023a, b, c; Sun et al., 2021) and boost passenger satisfaction (Moon et al., 2021). La et al. (2021) found that inflight entertainment facilities were the primary consideration for some passengers when choosing airlines, and, therefore, a quality inflight entertainment system could increase customer satisfaction. Park (2019) observed that digital services and inflight technologies, such as onboard entertainment and Wi-Fi, improved overall satisfaction and customer loyalty towards an airline. Earlier, Singaravelu and Amuthanayaki (2017) found that onboard entertainment and online ticket booking were significant for some Indian travellers. Therefore, they seemed to be attracted to airlines that provided such services. In sum, hi-tech innovation in the aviation industry is principally about using technologies to deliver services and carry out business activities. This includes using mobile CRM, cloud, blockchain technologies; big data, IoT, and robotics (see Fig. 6.1).

Hi-tech innovation in the aviation industry is not without challenges. For instance, although paper-based maintenance record keeping is still in the aviation industry (Abdallah & Fana, 2020), digital formats are being widely adopted by aviation bodies such as the International Civil Aviation Organisation (ICAO) and International Air Transport Association (IATA) initiate Guidance for Acceptance of Electronic Aircraft Maintenance Records (EAMR) (ICAO, 2017) and Paperless Aircraft Guideline (IATA, 2017) respectively. While hi-tech innovation improves efficiency and saves time, thereby reducing cost (Shiwakoti et al., 2022), there are practical aspects such as the establishment of shared regulatory standards, acceptance by stakeholders, adoption of universal requirements, such as lease agreements, and the harmonisation of technological solutions for adequate access and data (Scott, 2016). A report by the Centre for Digital Built Britain (CDBB) titled 'Challenges of Digitalisation in the Aerospace and Aviation Sectors' (Lamb, 2018) also described data interoperability, security, human factors in relation to acceptability, and social outcomes with regard to passenger experience, as some of the challenges of digital

Fig. 6.1 Digital technologies in the aviation industry. (Source: Authors)

transformation in the aviation industry. Establishing mutual regulatory standards is a crucial challenge for Digital Aircraft Operations and Maintenance (DAOM), mainly due to the different standards deployed by each regulatory authority (Abdallah & Fana, 2020). Nevertheless, the current status of hi-tech innovation in the aviation industry must be adequately explained to paint a picture of what is currently obtaining in the industry.

6.3 Methods and Materials

This qualitative study draws from the reports and websites of purposefully selected airlines worldwide. A total of 10 airlines globally were identified for analysis. These included Air Canada, JetBlue Airline, and WestJet (Americas); Air New Zealand (Oceania); *Emirates Airlines* and Southwest Airlines (Asia); Kenya Airways and Ethiopian Airways (Africa); and Ryanair Airlines and Lufthansa (Europe). The airlines were selected based on the availability of their annual reports and the content on their websites, and also the depth of information in the annual reports regarding hi-tech innovation. The authors' intuitions were used to judge the quality of the information in these airlines' annual reports and websites. In addition to reports and websites, the authors also utilised statements from senior officials of the selected airlines through interviews. These statements were accessed from various media outlets.

The study also draws from the personal observations of the authors as users of flights and as researchers who have published on similar topics before (Dube & Nhamo, 2019, 2020; Dube et al., 2020). In addition, the study also benefits from the review of existing global literature on airlines and the aviation industry. Content analysis was used to investigate what is in a text (Sheppard, 2021). This enabled the authors to inspect what is contained in the reports and websites of airlines and the existing literature so as to ensure that the texts formed the basis for drawing conclusions about hi-tech innovation in the selected airlines. Content analysis enabled the authors to integrate the information available in reports, and websites of the selected airlines to make conclusions about their state of digitalisation. Thematic data analysis—a method in which recurrent themes are identified from the available data—was used. Themes were generated deductively.

6.4 Findings

The analysis from the various data sources utilised in the article indicates that airlines have diverse and different levels of hi-tech innovation.

6.4.1 The JetBlue Airline

JetBlue Airlines has adopted technologies that personalise and broaden its business operations, streamline payments, and expand profit prospects. Among other innovations, it has integrated electric ground operations by converting its Ground Service Equipment to electric and maximising electric ground power and aircraft systems to minimise fuel use and reduce gas emissions (JetBlue Airways Corp, 2023). It was also among the first to introduce live TV and Wi-Fi onboard (OAG, 2022). The airline also adopted the 'customer 360 initiative', which uses Machine Learning and AI to identify customers, prepare for unanticipated events, and democratise data to attain efficiency (Gandharv, 2022). Using Snowflake's cloud-based data platform, the airline prides itself on a central dataset to make real-time and predictive decisions (JetBlue Airways Corp, 2023). JetBlue has also integrated AI and cutting-edge analytics to optimise operations, fuel usage, and flight logistics. Steve Taub, Head of Investments at JetBlue Ventures, posited that: "At JetBlue Technology Ventures (JTV), we think AI and specific use cases for autonomous flight technology have a real future in supporting the aviation industry's growth" (Taub, 2022).

JetBlue Airline has also integrated complex digital systems, which include an integrated reservations system, flight operations system, communications systems, website, upkeep systems, check-in kiosks, automated operation procedures, automated aircraft fleet control, and other assets and programmes that improve efficiency as well as collaboration with customers, staff, and suppliers (JetBlue Airways Corp, 2023). The airline is now heavily reliant on automated systems to conduct its business. This has reduced operation costs and increased customer demand (JetBlue Airways Corp, 2023).

6.4.2 WestJet

WestJet Airlines Ltd has also made significant strides in hi-tech integration in its operations, enabling it to remain competitive in recent years. The airline relies on computerised systems to conduct business and sustain its activities. These systems include flight control, computerised reservation, revenue management, communication, aircraft maintenance systems, airport kiosk terminals, and a website (WestJet, 2018). In mid-2018, WestJet introduced Juliet, a text-based, AI-powered chatbot. Juliet is available through Facebook Messenger and supports customers to easily manage numerous aspects of their journey, including destination discovery, flight booking, travel data, and many more straightforward aspects. This enables the airline's contact centre staff to focus on factors that automation may not resolve (WestJet, 2018).

WestJet Airline has also excelled in data connectivity and efficiency and has been recognised for the best inflight connectivity innovation using ZeroTouch technology (Bailey, 2021). ZeroTouch is a tool that enables media, software, and other material to be uploaded or offloaded onto the aircraft over the air without any technician import. Remote media loading enables technical operators to focus on critical maintenance and safety issues (WestJet, 2018). This has helped the airline to reduce data load times by 85%, eradicate time-consuming and labour-intensive practices, and provide customers with improved entertainment (OAG, 2022). ZeroTouch operates over current and future data links like Wi-Fi, cell modems, and Ku-band. According to Kris Valckx, Senior Fleet Engineer at WestJet, with ZeroTouch, the airline is "...able to quickly ascertain guest trends and pivot [its] product to match demand without overly stressing the manpower historically required to introduce new features to the fleet" (Bailey, 2021).

6.4.3 Air Canada

Air Canada is rapidly digitalising and integrating novel technologies into its operations. The airline rapidly adopts and integrates model digital technologies in its business operations. This has enabled it to increase profits and reduce operational costs (Air Canada, 2023). The technology systems used by the airline include those relating to its websites, automated customer sales and services, flight operations, loyalty programs, communications, accounting, business, and administrative systems. It uses AI and real-time data to make business decisions and predict aircraft capability (Saran, 2021). The airline was also among the first airlines to develop a voice service—Alexa; the first airline in Canada to provide in-seat Wi-Fi on selected flights and the first airline in Canada to introduce mobile booking on an Apple app (Air Canada, 2023).

The major outstanding digital innovation by the airline is its *digital identification technology mobile app*, which allows passengers to access multiple touchpoints along their journey without needing to show their boarding passports or government identity documents. This digital identification technology uses face measurements to create a numeric representation called a face print. The face print is then compared to verify a person's identity (Greenberg, 2022). Even before the COVID-19 pandemic in 2019, Air Canada implemented the SAP technology to fully integrate procumbent solutions, enabling staff to work remotely impeccably and resourcefully (Air Canada, 2023). Air Canada is also reputable for developing advanced technologies to generate less atmospheric carbon, reduce water waste, and decrease noise. Like all other airlines, Air Canada has not lagged in automating manual processes and reducing paperwork. This has been facilitated by SAP Ariba Invoice Management, which enables paperless invoicing.

6.4.4 Air New Zealand

According to OAG (2022), Air New Zealand has a long history of leading in innovative technologies that eventually leave a significant imprint in the airline industry. The airline has adopted and implemented several hi-tech innovation strategies meant to realise several goals ranging from security and customer satisfaction through smooth operations and services to climate change responsiveness and cost-effectiveness. The airline has implemented battery and hydrogen-powered electric aircraft to reduce operational costs and carbon emissions. Concerning extreme weather conditions, the airline has implemented flight software which uses complex data analytics to optimise flight routes once the aircraft is in flight (Air New Zealand, 2022). At Auckland, the airline has also installed a system which detects thunderstorm detection and provides notifications to the crew (Air New Zealand, 2022). The airline has also made tireless efforts to transform its existing aircraft technology towards renewable energy and to improve fuel efficiency.

Air New Zealand also introduced an AI chatbot called Oscar (Grigonis, 2018) and a mobile app used by customers in specific airport lounges to order coffee. The Oscar chatbot is designed to enable it to respond to some of the airline's regularly asked questions in a chat-like platform. This helps save time and offers a personalised experience for users. Together with Oscar, the airline's mobile app for iOS and Android also includes real-time data on flights, digital ticketing, digital boarding passes, and online flight booking. In 2015, the airline introduced Airband, a wearable wristband for unaccompanied children. This wristband enabled guardians to track the journey of their children. On its website, the airline has also listed a variety of hi-tech technologies which it uses to attain cybersecurity, meet environmental sustainability goals, use media for marketing, offer customer help and support, track and manage a fleet, manage data, and monitor and implement fleet maintenance, among others.

6.4.5 Emirates Airlines

The Emirates Airlines Group continues to embark on massive technological innovations to improve the travel experience and boost travel confidence. Since 2019, the airline has implemented novel digital tools that facilitate quick and confidential document verification, expanded biometrics, self-check-in, bag drop kiosks, and other contactless services at its Dubai centre (Emirates Group, 2023). Currently, the airline has over 30 biometric cameras in active operation at the exact centre, including check-in counters at the entries of its First and Business Class lounges and selected boarding gates (Emirates Group, 2023).

Emirates Group has also implemented frictionless boarding technology, which allows passengers to report delayed and damaged luggage (OAG, 2022). In partnership with its Dubai Airport stakeholders, including the immigration personnel, the airline activated Smart Gates, allowing authorised travellers to easily pass through passport control within seconds using new contactless technology (International Airport Review, 2021).

The Company has also increased investments in electric and hybrid ramps, ground support equipment (GSE) and forklifts and revamped the current GSE with digital tools. This is part of an ongoing fleet transformation approach to new sustainable technologies (Emirates Group, 2023). Emirates Airlines has also drawn from its partnership with the UAE's Aviation X-lab innovation incubator programme to introduce robot cleaning services to improve hygiene standards.

During the COVID-19 pandemic, the Emirates retired its hardcopy newsletter and increased staff assignation on digital channels such as Microsoft Teams Live (International Airport Review, 2021). The airline also introduced a 'OneCargo Management system' to digitalise its operations and maximise proficiencies across its cargo processes.

6.4.6 Southwest Airlines

With energy, health, and other types of crises on the rise and threatening air transport, digital technologies are providing novel ways to attain sustainable and cost-effective ways to deal with both intensifying ecological and financial pressures and optimising efficiency. Big Data and AI are the two digital levers that Southwest

Airlines uses to meet its carbon footprint targets and increase fuel efficiency (Southwest Airlines, 2023). Southwest Airlines Company is leading in using innovative technologies to monitor efficient and effective fuel consumption. Utilising Big Data and AI, the airline has developed eight predictive frameworks, which comprise time series algorithms and neural networks that enable its analytics system to generate up to 9600 fuel demand forecasts per month for over 100 airports (Ziemelis, 2022). The airline has also implemented novel technologies to enable carbon capture and sequestration.

In addition, the airline has installed onboard power ports on MAX aircraft to charge personal devices, such as mobile phones, at every seat. In 2022, the airline added a new automated payroll system and improved cyber controls (Southwest Airlines, 2023). Digital technologies have enabled the airline to make substantial enhancements in the areas related to flight and crew planning and scheduling; aircraft turn capacities; foundational operating systems; revenue management; technology infrastructure and cyber-security; and enhanced customer self-service offerings (Southwest Airlines, 2023). These tools and techniques include the Company's website and reservation system; flight dispatch and tracking systems; flight simulators; check-in kiosks; aircraft maintenance, planning, and record-keeping systems; telecommunications systems; human resources systems; and accounting systems (OAG, 2022).

6.4.7 Lufthansa Airlines

Another airline that has been making substantial progress in digitalisation and technology integration is Lufthansa Airlines. The airline is currently digitising all the features of its operations, digitally enhancing its portfolio with standalone digital offerings (Lufthansa Group, 2023). The OAG (2022) observed that Lufthansa had also excelled in utilising digital tools to develop a more 'customer-centric airline', which includes more digital passenger touchpoints and personalisation in its operations. The airline's technology relates to the following key areas: Alternative Fuels and Emissions, Aircraft Hardware, Digital Solutions and Processes, Waste and Circular Economy, and Mobility of Tomorrow and Beyond (Lufthansa Group, 2023).

Lufthansa Airlines also started regularly using the fuel-saving surface technology AeroSHARK (Lufthansa Group, 2023). This technology replicates the features of sharkskin, which is principally hydrodynamic, thus improving the aerodynamics of related aircraft features and using less fuel. In addition, the airline has introduced biometric identification technologies, contactless self-check points, bag drop kiosks, and contactless touchpoints (OAG, 2022). The first passenger aircraft with AeroSHARK was a Boeing 777, which started operating in October 2022.

6.4.8 Ryanair

Ryanair aims to be the number one airline. Thus, the airline is significantly investing in data analytics, software development, quality assurance and UX and design to meet this goal. The Company has already opened three labs in Ireland, Poland, and Spain for this purpose (Ryanair Group, 2023). Ryanair's reservations are made using new online technologies (OAG, 2022). The Ryanair Internet booking system allows potential customers to make reservations and book flights in real time through the Ryanair.com website. It also has a mobile app which customers can use to access its offers. The website and app also enable customers to add additional secondary products (e.g., luggage, priority boarding, and desired seating). The airline has also introduced a 'My Ryanair' account on its website, which makes flight booking. The 'My Ryanair' enables customers to securely store personal and payment details, considerably improving the booking process (Ryanair Group, 2023).

Ryanair has also developed a digital centre of excellence with complete flight simulators, a fixed base simulator, and a full-size Boeing 737NG training aircraft to enable pilot, engineering, and cabin crew training in Bergamo. The Boeing 737NGs are fitted with CFM 56-7B engines and have advanced CAT III Autoland capability, advanced traffic collision avoidance systems, and enhanced ground-proximity warning systems. The Boeing 737-8200 are fitted with CFM LEAP-1B engines which, together with the cutting-edge technology winglet and other aerodynamic advances, lower fuel consumption by around 16% on a per-seat basis when compared to the Boeing 737NGs. This also reduces operational noise emissions by approximately 40% (Ryanair Group, 2023).

Ryanair also launched significant technologies in response to the need for environmental sustainability in line with *Ryanair's Policy on Noise and Emissions*. Thus, the Company is investing in new, efficient aircraft and engine technologies that minimise the environmental impact of its activities. This has enabled Ryanair to be a leading competitor in matters of ecological efficiency (Ryanair Group, 2023).

6.4.9 Ethiopian Airways

Ethiopian Airlines is the largest aviation group in Africa. It has also been making significant efforts to adopt new technologies and digitalise its operations. In 2020, the airline upgraded its chatbot for domestic and international flight use (Ethiopian Airlines Group, 2022). The chatbot, nicknamed 'Lucy', has several self-service features that allow customers to process their travelling efficiently. The airline pioneered new digital aviation tools in Africa, including the first jet aeroplane and the first B767, B777-200LR, B787 Dreamliner, and 787-9 (Ethiopian Airlines Group, 2022). In 2017, the airline completely digitalised its business processes from aircraft maintenance, human resource management, and customer services to procurement and supply, becoming utterly paperless. A wide range of digital services, including online bookings, reservations, and self-check, is also available on the Company's website. The airline has also rolled out onboard Wi-Fi Internet connectivity using the latest satellite technology and launched a mobile app and chatbot-assisted shipment tracking service (Koigi, 2020). The mobile app is available for Android and iOS. It has improved convenience for customers of Ethiopian Cargo and Logistics Services through self-service features, including checking flight schedules, cargo tracking, and charter requests. The Ethiopian chatbot allows customers to access real-time shipment information using Messenger and Telegram (Ethiopian Airlines Group, 2022).

6.4.10 Kenyan Airways

Kenyan Airways has rapidly integrated new technologies to boost its operations and meet environmental sustainability commitments. Novel technologies are being used by the airline to control and monitor operations to close the gaps leading to high fuel use and aircraft maintenance costs (Kenya Airways Limited, 2021). There has also been the implementation of digital Flight Operations solutions. The Flight Operations suite incorporates numerous data sources like flight data, meteorological conditions, navigation, flight schedules, and other critical data to provide vital insights for airlines around fuel use and operations. The airline has partnered with GE Aviation to adopt its Flight Operations suite of digital products across its fleet of Boeing 737, 787 and Embraer E190 aircraft (Kenya Airways Limited, 2021). GE is the world's Digital Industrial Company, transforming the airline industry with software-defined technologies and solutions that are linked, reactive, and predictive.

Kenyan Airways also offers online booking, self-check-in services, and provides onboard WI-FI. It also utilises various novel technologies to attain cyber security, meet environmental sustainability goals, embark on online marketing, provide customer help and support, track and manage the fleet, and monitor and implement fleet maintenance.

6.5 Discussion

The review indicates that hi-tech innovation and integration in the aviation industry revolve around four broad themes: Operations Management, Flow monitoring and capacity management, Predictive and preventive solutions, and Digitisation costs and benefits, as summarised in Fig. 6.2.

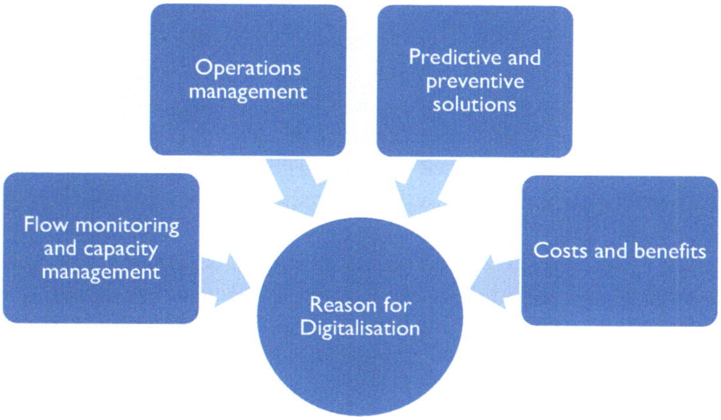

Fig. 6.2 Reasons for airline digitalisation and technology integration. (Source: Authors)

6.5.1 Operations Management

The review shows that one of the reasons for the rapid and widespread adoption and integration of AI technologies by global airlines is to improve operations. Although the selected airlines have different hi-tech adoption and integration levels, the review should show that AI technologies are generally part of their broader modern approach to information and communication expertise and the total management concept. All the reviewed airlines used the Internet and Cyber-Physical Systems in most of their operations. The Internet is used in collaboration with CPS—integrated physical and virtual systems that use computation, communication, and control systems—to facilitate access to real-time data and smart data management, analytics, and computational capability (Ndhlovu & Dube, 2023a, b, c). In collaborating with CPS, the Internet is used by airlines on instruments such as laptops, mobile phones, sensors, and actuators to expedite the integration of the physical and cyber worlds. This allows these airlines to integrate security, communication, computing, and entertainment in the boarding areas and on flights. This enables airline personnel to connect with passengers and assemble real-time data on current and potential customers. The comments on the websites of these selected airlines indicated senior personnel believed that using the IoT and CPS enables prompt, personalised, and localised services and precise appraisal of customer behaviours and likings. The implemented technologies included sensors, 3D scanners, cameras, and radio frequency identification instruments. All airlines reported having CPS on facilities to improve security and operations.

The adoption of smart technologies to improve services by airlines is consistent with existing empirical evidence, which shows that new technologies can potentially improve activities such as ground operations. It can also help airline management to develop collective decision-making and optimise their resources (Chakraborty et al., 2021; El Fkharany, 2022; Zaharia & Pietreanu, 2018). In addition, smart technologies can also provide the basis for the relationships between various parties within the aviation industry, establishing better collaboration between the airlines and airports, but also customers, employees, and investors.

Smart technologies are adopted by airlines to control their operations in real time and automate activities to satisfy customers and survive the competition. Technology integration enables airlines to reduce delays and risks and have a preventive slant to events. The selected airlines already use biometrics, Bluetooth, WI-FI, and other contactless technologies. These technologies can also be used to provide a systematic review of all activities, establish safety measures, and reduce operational risk. The airlines can also draw on real-time data inputs to visualise activities or AI for decision-making to increase effectiveness. Thus, smart technologies can be the basis of a holistic approach of airlines towards optimising their processes and activities and monitoring operations.

6.5.2 Flow Monitoring and Capacity Management

Customers' engagement in the contemporary world entails digital and collaborative use (Zaharia & Pietreanu, 2018). The whole intention of digitalisation and technology integration by airlines is to improve services, retain old customers, and attract new ones (Kenya Airways Limited, 2021; Lufthansa Group, 2023; Ryanair Group, 2023). Implementing self-services, such as self-check-in, self-checkout, and online booking, reservation, and payment, are all meant to save time for customers. This has the potential to solve queue challenges often experienced by customers at airports when passengers and luggage are processed manually.

Deploying automated customer flow solutions such as biometrics, image-based flow monitoring, and other contactless services can provide an ideal capacity process for airlines. Several reviewed airlines are already integrating smart technologies into E-luggage tags, E-menu, and Self-check-in kiosks, enabling customer data to be gathered and connected impeccably with relevant parties in real time.

In the contemporary business world, business success highly depends on collecting relevant data and using the data to inform decisions. All ten airlines in this study reported adopting or still working towards adopting Big Data analytics. Big data analytics are fast-replacing traditional data collection, processing, and analysis (Zeqiri et al., 2020). Big Data analytics enables enterprises to gather, process, and analyse vast consumer behaviours and preferences data through AI. It comprises internal data stored in central databases and external data collected from the Internet through sensors (Mhlanga et al., 2021). After processing and analysing data and placing it into categories according to their characteristics and type, airlines may be able to use the data to develop strategies to attract customers and improve their operations. Big data analytics could also enable these airlines to understand customer needs, behaviours, patterns of travel, and locations to offer personalised services.

Technologies are also crucial for ensuring safety and security at airports and not just within airlines. Security checkpoints are often the most problematic areas in the aviation industry, particularly congestion and overcrowding. For instance, in mid-December 2022, social media was flooded with images of angry passengers displaying the utter chaos at Delhi Airport's Terminal 3 in India. There were long lines at check-in and missed flights, with passengers ordered to carry only a single bag. The Chief Executive Officer of Delhi Airport was summoned to parliament on 15 December 2022 (IANS, 2022). Although it is often the immigration personnel that causes congestion and not necessarily the airlines themselves, the airline administration needs to explore solutions to increase processing capacity and ensure the optimisation of their facilities and not worsen the situation. Automatic security control systems made available by the advent of smart technologies represent a good solution. This should be implemented in collaboration with airports, increasing capacity by up to 30% when automated systems are implemented (Zaharia & Pietreanu, 2018).

The reviewed literature showed that the significant problems experienced by customers in the boarding area are demonstrated by the extended waiting time, poor Internet connection, and inadequate updates regarding travel changes or delays (Moon et al., 2021; Pereira, 2021; Razdan, 2022). On average, it has been observed that passengers spend about 30–50 min at the embarkation area (Coppens et al., 2018; Haygood, 2022). From a digital perspective, solutions for improving the passengers' experience and enhancing their satisfaction in boarding areas are related to options for entertainment, food ordering digital platforms, and self-service boarding systems. The implementation of automatic boarding systems has been observed as having the capacity to reduce boarding time by half (Hussain et al., 2022). Zaharia and Pietreanu (2018) found that about 400 passengers (3.3 s per passenger) can be embarked on a flight with automated systems. This will eliminate delays, increase the number of flights in and out of airports, and consequently boost customer satisfaction.

6.5.3 *Predictive and Preventive Solutions*

The review shows that the continued rise in the adoption and use of digital tools is meant to enable airlines to monitor, assess, and predict the fitness of aircraft infrastructure, systems, and components. These activities—which are often described as predictive maintenance, integrated vehicle health management, prognostics and health management (PHM), or aircraft health management (Scott et al., 2022)—make an integral part of a condition-based maintenance strategy that provides airlines with substantial benefits in terms of costs and time. In the ten airlines selected for the study, digital tools for predictive solutions were deployed for automated planning, management, and preventive actions related to security issues.

Aviation experts have observed that aircraft structural PHM rapidly tilts towards proactive condition-based maintenance tools (Hu et al., 2022). Structural PHM provides actionable data-driven decision-making, better operational proficiencies, concentrated maintenance costs, and better structural design (Scott et al., 2022). Decision-making support is an integral part of PHM. Using digital tools such as sensors, airlines can harness the PHM chain to detect and diagnose anomalies to inform decision-making. In this view, airlines can use digital technologies for operation optimisation, decrease jamming and delays, lowering costs, provide synergies amongst parties, and develop better business models. Digital tools can also be used for flight optimisation by identifying lucrative and efficient flight routes without causing substantial ecological damage. All the reviewed airlines, however, did not report robust and comprehensive systems that covered all aspects of preventive procedures. Investment is, therefore, required in this area.

6.5.4 Digitisation and Technology Integration Costs and Benefits

The reviewed airlines' primary motivation for digitalisation and technology integration is lower operational costs. Other aspects, such as environmental sustainability, only emerge as an appendage. In each of the annual reports of the selected airlines, the term 'cost reduction' repeatedly appeared, thus, signifying the importance of this aspect Air Canada, 2023; Kenya Airways Limited, 2021; Lufthansa Group, 2023; Ryanair Group, 2023). Scholarly literature also reiterates the diagnostic potential to reduce operational costs (Scott et al., 2022; Shiwakoti et al., 2022). Digital solutions have also enabled airlines to cut costs by providing automated non-remunerated technologies, digital marketing platforms, paperless invoicing, self-check-in and out, and remote reservations. Airlines can generate more personalised and distinctive experiences at reduced costs through AI and robots. Service robots enable them to offer services at any time of the day. AI and intelligent robots have been found to be more affordable, faster, and even more reliable (Zeqiri et al., 2020). These tools can also decrease costs as human personnel is only utilised in areas where machines cannot offer services. Solar-powered technologies have also allowed airlines to save on electric and fuel energy costs. Through technology-based human resources systems, airlines can manage staff behaviour and the work situation more efficiently to reduce turnover and associated costs (Kabadayi et al., 2019). Paperless documentation is not only cost-effective, but it is also environmentally responsible. Instead of using painted walls or physical billboards, through augmented reality (AR), airlines could utilise a blend of real and virtual objects in a natural environment, synchronise real and virtual objects, and interface in 3D and in real-time. Projection-based AR-enabled airlines to project artificial light on the surface of physical things to advertise.

6.6 Conclusion

The article provides a general description of the state of digitalisation and technology integration in ten selected airlines worldwide. Focus was placed on the reasons for digitalisation and technology integration. The study found different levels of digitalisation within airlines, with some airlines outshining others considerably. Overall, the study found that digital tools were being rapidly adopted and integrated into airlines for four significant reasons: to improve operations management, improve flow monitoring and capacity management, enable predictive and preventive solutions, and reduce operational costs.

A significant limitation of the study is that it does not qualitatively explore the reasons behind differentiated digitalisation by airlines. The authors also observed impressive digitalisation in airlines operating in some global regions than others. It is vital, therefore, that future studies consider these research gaps. It is also essential to explore the attitudes of both staff and customers towards digitalisation. Satisfying these research gaps might also necessitate a different research methodology from the one used in this study.

References

Abdallah, A. A., & Fana, I. (2020). Emerging challenges of digital aircraft operations and maintenance: A knowledge management perspective. In *9th international conference on through-life engineering service 3–4 November 2020, Cranfield UK*.

Afaq, A., Gaur, L., Singh, G., & Dhir, A. (2021). COVID-19: Transforming air passengers' behaviour and reshaping their expectations towards the airline industry. *Tourism Recreation Research*, 1–9.

Air Canada. (2023, February). *2022 management's discussion and analysis of results of operations and financial condition*. Air Canada. https://aircanada.investorroom.com/image/Q4+2022+Air+Canada+MD%26A+-+English.pdf

Air New Zealand. (2022). *Annual financial results*. Air New Zealand. https://www.annualreports.co.nz/wp-content/uploads/2022/10/AIR.2021.pdf

Airasia. (2022). Available online: https://www.airasia.com/aa/about-us/en/gb/scan2fly.html. Accessed 19 Mar 2023.

Amankwah-Amoah, J. (2021). COVID-19 pandemic and innovation activities in the global airline industry: A review. *Environment International, 156*, 106719.

Apăvăloaie, E. (2014). The impact of the internet on the business environment. *Procedia Economics and Finance, 15*, 951–958.

Bailey, J. (2021, December 09). How WestJet is cutting data load times by 85%. *Airline News*. http://simpleflying.com/wetjet-panasonic-zerotouch/. Accessed 30 Mar 2023.

Chakraborty, S., Chakravorty, T., & Bhatt, V. (2021). IoT and AI driven sustainable practices in airlines as enabler of passenger confidence, satisfaction and positive WOM: AI and IoT driven sustainable practice in airlines. In *Proceedings of the international conference on artificial intelligence and smart systems (ICAIS), Coimbatore, India, 25–27 march 2021* (pp. 1421–1425).

Chihwai, P., Ndhlovu, E., & Dube, K. (2023). Exploring trends of South African tourist accommodation occupancy during COVID-19. In K. Dube, I. Mensah, & L. Chapungu (Eds.), *COVID-19, tourist destinations and prospects for recovery*. Springer.

Cohen-Almagor, R. (2011). Internet history. *International Journal of Technoethics, 2*(2), 45–64.

Coppens, J., Dangal, S., Vendel, M., Anjani, S., Akkerman, S., Hiemstra-van Mastrigt, S., & Vink, P. (2018). Improving airplane boarding time: A review, a field study and an experiment with a new way of hand luggage stowing. *International Journal of Aviation, Aeronautics, and Aerospace, 5*(2), 1–12.

Deepika, K., Tilekya, V., Mamatha, J., & Subetha, T. (2020). Jollity Chatbot-a contextual AI assistant. In *Proceedings of the third international conference on smart systems and inventive technology (ICSSIT), Tirunelveli, India, 20–22 August 2020* (pp. 1196–1200).

Delgado-Mohatar, O., Fierrez, J., Tolosana, R., & Vera-Rodriguez, R. (2020). Blockchain meets biometrics: Concepts, application to template protection, and trends. *arXiv*, arXiv:2003.09262.

Delta. (2021). *Delta reveals first-ever dedicated TSA Precheck Lobby, Bag Drop*. Available online: https://news.delta.com/delta-reveals-first-ever-dedicated-tsa-precheckr-lobby-bag-drop. Accessed 27 Mar 2023.

Dube, K., & Nhamo, G. (2019). Climate change and the aviation sector: A focus on the Victoria Falls tourism route. *Environmental Development, 29*, 5–15.

Dube, K., & Nhamo, G. (2020). Major global aircraft manufacturers and emerging responses to the SDGs agenda. In G. Nhamo & O.G.M.V. (Eds.), *Scaling up SDGs implementation emerging cases from state, development and private sectors* (pp. 99–113). Springer International Publishing, Springer Nature Switzerland AG.

Dube, K., Nhamo, G., & Chikodzi, D. (2020). COVID-19 cripples global restaurant and hospitality industry. *Current Issues in Tourism*, 1–4.

Dube, K., Nhamo, G., & Chikodzi, D. (2021). COVID-19 pandemic and prospects for recovery of the global aviation industry. *Journal of Air Transport Management, 92*, 102022.

El Fkharany, M. H. (2022). Applying biometric Technology for Enhancing Airports' efficiency: A technology acceptance model as an assessing tool. *International Journal of Tourism and Hospitality, 3*(2), 104–117.

Emirates Group. (2023). *Emirates group annual report 2021–2022*. Emirates Group.

Ethiopian Airlines Group. (2022). *Annual report 2020–2021*. Ethiopian Airlines Group. https://corporate.ethiopianairlines.com/docs/default-source/annual-performance-reports/annual-report-2020-2021.pdf?sfvrsn=60ea08e4_6

Gandharv, K. (2022, June 29). *How JetBlue uses data to reach new heights*. Staff articles. JetBlue Uses Data To Reach New Heights (cxmtoday.com).

Greenberg, K. E. (2022, May 31). *Air Canada flies into re digital stratosphere*. BrandPost. https://www.cio.com/article/350369/air-canada-flies-into-the-digital-stratosphere.html

Grigonis, H. K. (2018, April 4). *AI helps Oscar the Chabot answer 75 percent of travel questions*. Digitaltrends. https://www.digitaltrends.com/mobile/oscar-air-new-zealand-chatbot/

Haas, H. (2018). LiFi is a paradigm-shifting 5G technology. *Reviews in Physics, 3*, 26–31.

Haygood, J. (2022, August 30). *When do plane start and stop boarding*. SkyTouch. https://www.skytough.com/post/when-do-planes-start-stop-boarding#:~:text=With%20the%20average%20boarding%20time,not%20reopen%20them%20for%20you. Accessed 29 Mar 2023.

Hu, Y., Miao, X., Si, Y., Pan, E., & Zio, E. (2022). Prognostics and health management: A review from the perspectives of design, development and decision. *Reliability Engineering and System Safety, 217*, 108063.

Hussain, S., Jabbar, H., Shehzad, H., & Aden, A. A. (2022). How automation revolution airport operation and capacity in post covid-19 period. A case study on Ankara Esenboga airport. *International Journal of Advance Research and Innovative Ideas in Education, 8*(3), 2857–2866.

IANS. (2022, December 13). *Delhi Airport CEO summoned by parliament panel on Dec 15 to discuss congestion issues*. World Asia. https://gulfnews.com/world/asia/india/delhi-airport-ceo-summoned-by-parliamentary-panel-on-dec-15-to-discuss-congestion-issues-1.92675513

International Air Transport Association, (IATA). (2017, November 1). *Guidance material for the implementation of digital aircraft operations in technical operations (PAO: TO)*. Available: https://www.iata.org/en/programs/opsinfra/digital-aircraft-operations/. Accessed 17 Mar 2023.

International Airport Review. (2021, September 13). *Emirates accelerates digital innovation for passengers*. https://www.internationalairportreview.com/news/163755/emirates-accelerates-digital-innovation-for-passengers/. Accessed 10 Apr 2023.

International Civil Aviation Organization (ICAO). (2017). *Guidance for acceptance of Electronic Aircraft Maintenance Records (EAMR)*, 11 2017. Available: https://www.icao.int/safety/air-navigation/OPS/airworthiness/Pages/EAMR.aspx. Accessed 27 Mar 2023.

JetBlue Airways Corp. (2023). *JetBlue airways corporation – 10K – Annual report*. JetBlue Airways Corp. https://fintel.io/doc/sec-jetblue-airways-corp-1158463-10k-2023-february-27-19415-2068

Kabadayi, S., Ali, F., Choi, H., Joosten, H., & Lu, C. (2019). Smart service experience in hospitality and tourism services. *Journal of Service Management*. https://doi.org/10.1108/josm11-2018-0377

Kenya Airways Limited. (2021). *Annual report*. https://africanfinancials.com/document/ke-kq-2021-ar-00/

Koigi, B. (2020, May 6). *Ethiopian Airlines launches innovations to elevate cargo customers' experience*. Africa Tech. https://africabusinesscommunities.com/tech/tech-news/ethiopian-airlines-launches-innovations-to-elevate-cargo-customers-experience/. Accessed 12 Apr 2023.

Kuisma, N. (2018). *Digitalization and its impact on commercial aviation*. Bachelor's Thesis. Aalto University, Espoo, Finland. Available online: https://aaltodoc.aalto.fi/handle/123456789/32996. Accessed 23 Mar 2023.

La, J., Bil, C., & Heiets, I. (2021). Impact of digital technologies on airline operations. *Transportation Research Procedia, 56*, 63–70.

Lamb, K. (2018). *Challenges of digitalisation in the aerospace and aviation sectors*. Centre for Digital Built Britain, University of Cambridge.

Lufthansa Group. (2023). *Annual report 2022*. Lufthansa Group. https://investor-relations.lufthansagroup.com/fileadmin/downloads/en/financial-reports/annual-reports/LH-AR-2022-e.pdf

Manyika, J., Chui, M., Bisson, P., Woetzel, J., Dobbs, R., Bughin, J., & Aharon, D. (2015). *Unlocking the potential of the Internet of Things*. McKinsey Global Institute. Available online: https://www.mckinsey.com/business-functions/mckinsey-digital/our-insights/the-internet-of-things-the-value-of-digitizing-the-physical-world. Accessed 25 Mar 2023.

Mhlanga, D., Ndhlovu, E., & Hofisi, C. (2021). Assessment of the 4IR challenges of radical innovation in service delivery in Africa. *Journal of Public Administration, 56*(4.1), 1002–1017.

Molchanova, K. (2020). A review of digital technologies in aviation industry. *Logistics and Transport, 47–48*, 69–77.

Moon, H. G., Lho, H. L., & Han, H. (2021). Self-check-in kiosk quality and airline non-contact service maximization: How to win air traveler satisfaction and loyalty in the post-pandemic world? *Journal of Travel & Tourism Marketing, 38*, 383–398.

Ndhlovu, E., & Dube, K. (2023a). Impact of COVID-19 on tourism in South Africa: Building resilience and prospects for recovery. In K. Dube, G. Nhamo, & M. P. Swart (Eds.), *COVID-19, tourist destinations and prospects for recovery volume one: A global perspective* (pp. 81–100). Springer Nature Switzerland.

Ndhlovu, E., & Dube, K. (2023b). Challenges of radical technological transition in the restaurant industry within developing countries. *African Journal of Hospitality, Tourism and Leisure, 12*(1), 156–170.

Ndhlovu, E., & Dube, K. (2023c). Restaurants industry recovery challenges from COVID-19 amidst the Russo-Ukrainian war. In K. Dube, G. Nhamo, & M. P. Swart (Eds.), *COVID-19, tourist destinations and prospects for recovery volume one: A global perspective* (pp. 337–356). Springer Nature Switzerland.

Noviantoro, T., & Huang, J. P. (2021). Investigating airline passenger satisfaction: Data mining method. *Research in Transportation Business and Management, 43*, 100726.

Official Airline Guide's Database (OAG). (2022, November 9). *Spotlight on airline tech innovators*. Future of Travel. https://www.oag.com/blog/spotlight-on-airline-innovators?utm_medium=email&_hsmi=233180523&_hsenc=p2ANqtz%2D%2DH0FLHOLVbU4x1

KclzPJmDTph4hME6YhzV76cBwxzcowVVK6iAW_oVPlczug7ZSm4YCHMYQXJ_GZ6qPdJjb0i3cfj7xg&utm_content=233180523&utm_source=hs_email. Accessed 29 Mar 2023.

Olaganathan, R. (2021). Impact of COVID-19 on airline industry and strategic plan for its recovery with special reference to data analytics technology. *Global Journal of Engineering and Technology Advances, 7*, 33.

Park, E. (2019). The role of satisfaction on customer reuse to airline services: An application of Big Data approaches. *Journal of Retailing and Consumer Services, 47*, 370–374.

Pereira, F. C. (2021). *The impact of the COVID 19 pandemic on European airlines' passenger satisfaction*. Master's thesis, University Institute of Lisbon. Available online: https://repositorio.iscte-iul.pt/handle/10071/23516. Accessed 17 Mar 2023.

Razdan, N. (2022, December 22). Delhi Airport overcrowding: What led to the massive congestion and chaos? *Gulf News*. http://gulfnews.com/opinion/op-eds/delhi-airport-overcrowding-what-led-to-the-massive-congestion-and-chaos-1.92828199. Accessed 21 Mar 2023.

Rotsidis, A., Theodorou, A., Bryson, J. J., & Wortham, R. H. (2019). Improving robot transparency: An investigation with mobile augmented reality. In *Proceedings of the 28th IEEE international conference on Robot and Human Interactive Communication (RO-MAN), New Delhi, India, 14–18 October 2019* (pp. 1–8).

Ryanair Group. (2023). *Annual report 2022*. Ryanair Group. https://investor.ryanair.com/wp-content/uploads/2022/07/Ryanair-2022-Annual-Report.pdf

Saran, C. (2021, October 7). *How data drives Air Canada's cargo business: Data-driven insights and artificial intelligence are helping Air Canada to find a new business path as it expands its cargo business following the pandemic*. CombuterWeekly.Com. http://www.computerweekly.com/news/252507721/how-data-air-canadas-cargo-business. Accessed 29 Mar 2023.

Scott, T. (2016). *Digital aircraft and angine lease returns* (White paper). https://www.aircraftit.com/articles/digital-aircraft-and-engine-lease-returns/. Accessed 20 Mar 2023.

Scott, M. J., Verhagen, W. J. C., Bieber, M. T., & Marzocca, P. (2022). A systematic literature review of predictive maintenance for defence fixed-wing aircraft sustainment and operations. *Sensors, 22*, 7070. https://doi.org/10.3390/s22187070

Serrano, F., & Kazda, A. (2020). The future of airports post-COVID-19. *Journal of Air Transport Management, 89*, 101900.

Sheppard, V. (2021). *Research methods for the social sciences: An introduction*. Saylor Academy.

Shiwakoti, N., Hu, Q., Pang, M. K., Cheung, T. M., Xu, Z., & Jiang, H. (2022). Passengers' perceptions and satisfaction with digital technology adopted by airlines during COVID-19 pandemic. *Future Transportation, 2*, 988–1009.

Singaravelu, K., & Amuthanayaki, V. (2017). A study on service quality and passenger satisfaction on Indian airlines. *Journal of Commerce and Trade, 12*, 106–115.

Southwest Airlines. (2023). *Southwest Airlines Co. 2022 annual report to shareholders*. Southwest Airlines. https://www.southwestairlinesinvestorrelations.com/~/media/Files/S/Southwest-IR/LUV_2022_Annual%20Report.pdf.

Suk, M., & Kim, W. (2021). COVID-19 and the airline industry: Crisis management and resilience. *Tourism Review, 76*, 984–998.

Sun, X., Wandelt, S., & Zhang, A. (2021). Technological and educational challenges towards pandemic-resilient aviation. *Transport Policy, 114*, 104–115.

Tan, J. H., & Masood, T. (2021). *Adoption of industry 4.0 technologies in Airports—A systematic literature review*. Available online: https://arxiv.org/abs/2112.14333. Accessed 23 Mar 2023.

Taub, S. (2022, June 20). Is there a future for autonomous flight? *JetBlue Ventures' Insights*. https://medium.com/jetblue-ventures-insights/the-future-of-autonomus-flight-aefde61f2ec

Tyagi, S., & Lodewijks, G. (2022). Optimisation of check-in process focused on passenger perception for using self-service technologies at airport in Australia. *Journal of Airline and Airport Management, 12*, 1–14.

West, D. M. (2016). *How 5G technology enables the health internet of things*. Centre for Technology Innovation at Brookings.

WestJet. (2018). *2018 annual report*. WestJet Airlines Ltd.

Zaharia, S. E., & Pietreanu, C. V. (2018). Challenges in airport digital transformation. *Transportation Research Procedia, 35*, 90–99.

Zeqiri, A., Dahmani, M., & Youssef, A. B. (2020). Digitalization of the tourism industry: What are the impacts of the new wave of technologies. *Balkan Economic Review, 2*, 63–82.

Ziemelis, G. (2022, August 9). *Crisis is coming: Big Data and AI could save aviation*. http://www.aerotime.aero/articles/32217-big-data-ai-aviation-fuel-crisis-ziemelis. Accessed 10 Apr 2023.

Part III
Barriers to and Challenges of Technology Adoption in Tourism Businesses

Chapter 7
The Opportunities and Challenges of Harnessing Digital Technologies in Wildlife Tourism Resources Conservation in Zimbabwe

Zibanai Zhou and Tendai Chibaya

7.1 Introduction

Wildlife tourism has huge socioeconomic significance in emerging economies such as Zimbabwe because of numerous benefits at the community and national level through employment creation and foreign currency earnings prowess (Kishnani & Sharma, 2022; Mbaiwa, 2017; Zhou, 2023). Based on this, governments, conservationists, and the non-governmental sector continue to set aside financial resources to enhance the conservation of wildlife tourism resources. However, despite such enormous budgetary support commitment, the illicit pillaging of natural resources continues unabated (Ramutsindela, 2016; Stoett, 2019). Consequently, sustainable wildlife tourism resource conservation paradigms dominate policy discourses at a global scale (Gandiwa et al., 2014; Muboko et al., 2016). The tourism industry in developing countries hinges on sustainable use and preservation of finite wildlife resources (Zhou, 2019a); without which, most wildlife resources would become extinct in accordance with the doctrine of the tragedy of the commons. For the past two decades, stakeholders have been scaling up efforts towards the conservation of wildlife resources, despite increasing challenges (Buscher & Ramutsindela, 2015; Gandiwa et al., 2014). Widespread poaching, veld fires, and climate change-induced deforestation are some of the challenges which continue to derail conservation initiatives. To this end, it is pertinent to continuously review contemporary wildlife resource conservation initiatives with a view to adapt and strengthen their effectiveness in addressing evolving challenges.

Z. Zhou (✉)
Midlands State University, Gweru, Zimbabwe

T. Chibaya
Great Zimbabwe University, Masvingo, Zimbabwe

E. Ndhlovu et al. (eds.), *Tourism and Hospitality for Sustainable Development*,
https://doi.org/10.1007/978-3-031-63069-9_7

The complexity and magnitude of threats to biodiversity and conservation of wildlife resources in Africa in general and Zimbabwe in particular are a cause of global concern (Anderson & Jooste, 2014; Matungwa & Wawa, 2021). This is based on growing concerns that illicit trafficking of wildlife resources in African countries is the main source of funding for many political conflicts across the globe (Mareto, 2018; Stoett, 2019). Consequently, this brings forth an interesting dimension to view the importance of wildlife resources conservation not only from the usual world view of supporting biodiversity, socio-economic livelihoods, and tourism but also from maintaining regional and global peace perspective. To this end, Ceballos et al. (2017), Schwartz et al. (2018), and Travers et al. (2019) emphasised a need to intensify global conservation efforts, coupled with requisite decision-making support frameworks to prevent further loss.

Wildlife resources support a thriving wildlife tourism sector worth billions of dollars, which also supports the livelihoods of downstream communities (Agyeman et al., 2019; Ariya et al., 2017; Zhou, 2019b). Wildlife tourism resource conservation programmes are steeped in the protectionist paradigm. This is partly a source of problems bedevilling wildlife conservation initiatives given that the primary concern of a protectionist paradigm is preservation at the expense of host communities (Gandiwa et al., 2014). Most local communities in developing countries lack basic necessities which cannot be reconciled with conservation (Mbaiwa, 2017). This contradicts official narratives which portray local communities as the traditional and rightful custodians of wildlife tourism resources. Most often, communities suffer from exclusion in tourism development processes (Saar et al., 2021), and are rarely listened to on matters regarding the use of wildlife tourism resources. Exclusive conservation has always been a source of conflict between conservationists and local communities. However, local communities have immense indigenous knowledge systems around the conservation of wildlife resources, which remain untapped.

Generally, the loss of wildlife biodiversity worldwide has increased significantly and has become a cause of concern (Brondizio et al., 2019; Johnson et al., 2019; Quesada, 2019). The unprecedented surge in poaching of endangered species, and large-scale loss of biodiversity induced by exogenous factors has put wildlife tourism resource conservation programmes' efficacy under the spotlight. One way through which the conservation of wildlife tourism resources can be broadened is through the use of Information and Communication Technologies (ICTs) and digital technologies. Wealthier countries mostly in North America and Western Europe are on the front foot in harnessing ICTs in wildlife tourism resource conservation programmes as compared to developing countries. Digital technology refers to the use of information and communication technologies and related software to capture, store, synthesise, and retrieve information. The study focuses on digital technologies encompassing the use of drones, artificial intelligence (AI), remote sensing, geo-spatial mapping, and cameras, among others. It is pertinent that southern Africa in general and Zimbabwe in particular explore innovative ICT-based conservation approaches and other emerging digital technology tools in order to broaden and enhance the conservation of wildlife tourism resources. While there is significant literature on wildlife resource conservation, notably Chinyandura et al. (2021), Gandiwa et al. (2014), Muboko et al. (2016), and Muposhi et al. (2016), studies exclusively focusing on the application of digital technologies to wildlife tourism resource conservation beyond poaching menace remain scant. The present study expands its focus to threats emanating from veld fires and climate change.

In the present study wildlife tourism resources refer to faunal and floral species-mammals like elephants, rhinoceros, hippos, buffaloes, other endangered species, vegetation, rangelands, forests, water bodies, and aquatic life; which constitute the bedrock of Zimbabwe's tourism industry (Zhou, 2018). There seems to be a paradox in the broader wildlife conservation sector. This stems from the reality that the successful implementation of conservation initiatives has had unintended consequences. In cases where wildlife conservation initiatives have been successful, wild animals have increased exponentially to unsustainable levels from an ecological perspective. In contrast, in instances where the conservation initiatives have been ineffective poaching threatens to decimate the faunal species. Many wildlife tourism resource conservation programmes have been rolled out in partnership with local communities; however, the success of such initiatives overall remains questionable. Additionally, stakeholder collaborations in wildlife conservation have been organically disjointed emanating from diverse and competing interests, dominance of certain actors over others, unequal roles and relationships, and inequitable benefits sharing, as well as mistrust issues. To this end, the debate around the conservation of wildlife tourism resources continues to evolve in search of a lasting solution, which however remains elusive. Research suggests that the exploitation of wildlife resources within a protectionist framework largely remains exclusionary to local communities (Gandiwa et al., 2014; Schmidt & Uriely, 2019); in favour of remote actors. This view is strengthened by the controversy that surrounds the opaque wildlife tourism resource conservation schemes, which are bankrolled and controlled by external actors at the expense of local communities. The state of affairs has been perpetuated by the general lack of financial resources at the central government level to support wildlife tourism conservation programmes, thus creating a vacuum for the so-called development partners to step in. With the advent of ICTs and digital technologies, there is a need to identify specific wildlife tourism resource conservation aspects where digital technologies can be deployed to strengthen and broaden the initiatives.

Despite documented historical successes of wildlife conservation initiatives, in the past two decades there has been a resurgence of tremendous pressure on wildlife conservation initiatives as evidenced by loss of biodiversity, widespread decline in animal population, and dwindling of endangered species due to emerging threats and other exogenous factors (Pimms et al., 2014). To this end, the effectiveness of the conservation initiatives has come under scrutiny. However, the situation is not hopeless, given the advent of ICTs and related digital technologies, which can be leveraged to strengthen wildlife conservation programmes.

Currently, scholarly research on the mediating role of digital technologies in wildlife tourism resource conservation in Zimbabwe's national parks is limited. Therefore, the research aims to examine the opportunities and challenges of harnessing digital technologies to conserve wildlife tourism resources. The research also assesses the effectiveness of particular digital technologies in diverse aspects of wildlife resource conservation, hence assisting the wildlife tourism resource conservation authorities to determine specific techniques and tools which are worth investing in and where. The study seeks to answer the following central research questions: (i) To what extent has wildlife tourism resource conservation been transformed after the adoption of digital technologies at practical level? (ii) Which specific digital technologies have been adopted by national parks to conserve wildlife resources? (iii) What opportunities are available to deploy digital technologies to conserve wildlife tourism resources and on what aspects of the wildlife tourism resource conservation value chain? (iv) What are the challenges faced by wildlife conservation authorities in adopting and deploying digital technologies in the Zimbabwean national parks context? (v) Has the adoption and use of ICTs and digital technology tools improved stakeholder collaboration in wildlife tourism resource conservation? Tackling the research questions provides insight into the effectiveness and limitations of digital technologies in wildlife tourism conservation and against poaching, veld fires, and climate change. The study is organised as follows: following this introduction, a literature review is provided; thereafter, the methodology is explained, the findings and discussion section then follows; and the conclusion is then given.

7.2 Literature Review

7.2.1 Wildlife Tourism Resource Conservation Discourses

Conservation of wildlife is largely driven by local needs and international obligations. Essentially, wildlife resource conservation is inspired by the realisation that its unregulated use can lead to the extinction of some wildlife species (Gandiwa et al., 2014; Wyatt, 2013). The overwhelming exploitation of wildlife species through hunting activities prompted people to call for the protection of the remaining wildlife populations through various conservation initiatives (Adams, 2013). The significance of wildlife resources is multi-pronged: they support socio-economic livelihoods, they are the bedrock of tourism; a source of natural heritage and a symbol of national pride. Wildlife management and conservation involve a lot of activities including law enforcement, fire management, water supply, and population monitoring (Bothma & Du-Toit, 2016), which require funding. Lindsey et al. (2016) caution that there is still a major funding shortfall for conservation in Africa. The dominant narrative in wildlife tourism resource conservation is that local communities continue to play a peripheral role or participate as junior partners given the reality that most conservation programmes are externally funded (Gandiwa et al., 2014; Muboko et al., 2016). Wildlife conservation programmes face funding cuts (Lindsey et al., 2021); and the decrease in ecotourism activities further reduces revenue generation and financial resources meant for wildlife conservation (Neupane, 2020; Waikhata, 2020). In support, Spenceley et al. (2021) assert that COVID-19 had a significant impact on wildlife management and curtailed funds to enforce poaching regulations. Wildlife trafficking can undercut sustainable development investments (Agu & Gore, 2020), and poses significant threats to international conservation efforts. Africa, although megadiverse, remains a focal point for global wildlife trafficking research, policy, and practice. Illegal wildlife trade has a direct bearing on the efforts of African governments to achieve their Sustainable Development Goals [SDGs] (Vandome & Vines, 2018).

Broadly, wildlife conservation has been largely funded by trophy hunting activities, ecotourism, government budgetary support, and development partners in the form of NGOs. Muposhi et al. (2016) outline that trophy hunting has been integral in supporting conservation financing in Zimbabwe. Trophy hunting has been established as a conservation tool supported by ecological theory and sustainable use principles (Mahoney & Jackson, 2013); however, it has been riddled with controversy. As an example, the illegal killing of Cecil, a radio-collared lion (*Panthera leo*) by an American trophy hunting client in western Zimbabwe re-ignited the controversy around trophy hunting (Gross, 2015). Muposhi et al. (2016) acknowledge that there is a perpetual debate, polarisation, and lack of consensus among conservation NGOs, some African governments, and animal rights and welfare groups over the acceptability and effectiveness of trophy hunting as a conservation tool.

Wildlife conservation and protection of habitat is a costly undertaking as this land use option is constantly under threat due to the continued human population increase. The main drivers of wildlife species loss in most African countries include the growing human population, illegal wildlife poaching, and trade (Atkins et al., 2018; Herbig & Minnaar, 2019); veld fires, habitat loss, and competing priorities on land use options (Lindsey et al., 2016). According to Haas and Ferreira (2018), wildlife poaching and trafficking are the most noticeable and widespread phenomena that compromise the integrity of biodiversity at the global level. There has been a significant decline in the elephant population in some parts of Africa attributed to overharvesting, habitat loss, and droughts (Craigie et al., 2010). In the same vein, the increasing illegal hunting activities in Zimbabwe's national parks are a worrying and significant threat to conservation efforts (Muboko et al., 2016; Gandiwa et al., 2014). The illegal hunting through poisoning has resulted in a global concern on conservation initiatives of elephants and rhinoceroses in Zimbabwe (Muboko et al., 2014, 2016). This puts into perspective the enormity of the war against poaching. According to Dube (2021), crisis times such as the COVID-19 pandemic heighten threats to wildlife conservation because of the inability of wildlife patrollers to provide protection services to wildlife in national parks necessitated by lockdowns. Neupane (2020) observed a surge in poaching in areas where communities resort to illegal hunting of wildlife as a means of survival under stressful environments such as the COVID-19 pandemic. Inadequate conservation budget and political interference were also found to be driving forces for increased poaching and wildlife loss. Illegal wildlife poaching is a growing crisis across the globe and undermines the conservation of wildlife resources, economic livelihoods, and tourism, threatening to completely wipe out wildlife, such as elephants, rhinos, and other endangered species, which are apparently the backbone of the African tourism sector. The hidden costs of poaching in the wildlife-rich African countries, apart from the decline and extinction of some wildlife species include a reduction in foreign currency earnings, destruction of natural heritage resources, and flight of foreign direct investment in the wildlife tourism sector (Agu & Gore, 2020).

The evolution of wildlife conservation in Zimbabwe is divided into three phases, namely, precolonial (before 1890), colonial (1890–1979), and post-colonial (1980 to date) (Muposhi et al., 2016). According to Muboko and Murindagomo (2014), wildlife protection is steeped in fortress or protectionist conservation paradigm, in which the collective access of indigenous communities has been transformed into a highly restrictive and segregatory policy. The fortress ideology resulted in the dispossession and displacement of indigenous people from their ancestral land and alienated them from the resources they were much dependent upon. Fynn and Kolawole (2020) posit that local people are pushed to the periphery of wildlife-protected areas where they seldom benefit from conservation, yet they are confronted with the serious challenge of having to contend with conflict with wildlife. Conservation actors, policymakers, donors, and communities should move beyond the premise of the fortress conservation paradigm, which assumes conflictual relationships between rural communities and wildlife. The postcolonial era experienced rapid policy advances towards inclusiveness in natural resources management by ensuring that communities living with wildlife resources ought to benefit from managing such resources (Gandiwa et al., 2014).

A variety of techniques have been used to preserve wildlife resources. The widely used traditional anti-poaching techniques encompass patrols, intelligence-led, conservation education, de-snaring, and sniffer dogs (Gandiwa et al., 2014). However, political interference, budget constraints, and inadequate manpower have been cited as the main obstacles hindering successful wildlife conservation. An alternative narrative suggests a need to combat poverty, through the establishment of economic opportunities and the full involvement of communities in decision-making in order to save wildlife. This view is premised on the understanding that dual approaches to combat further loss of wildlife resources are critical, for example, establishing strong institutions coupled with poverty eradication (Mbaiwa, 2017). Wildlife species can become extinct without drastic intervention by protection authorities. For example, Goldenberg (2013) and Nellemann et al. (2014) assert that poachers killed 35,000 elephants in 2012, which translates to 96 elephants being killed each day. The World Wildlife Fund (2012) revealed that in South Africa rhino poaching increased fivefold with 574 rhinos killed in 2013. This shows the ever-increasing complexity, global span, and need for new measures to counter animal poaching. Poaching threatens the safety of rural communities, undermines or reverses decades of conservation and development, drains source countries of their natural heritage, and jeopardises revenue from activities such as wildlife tourism. The realisation that the current wildlife conservation initiatives in southern Africa, including Zimbabwe, cannot cope with evolving threats on wildlife resources makes a case to explore other contemporary means of protecting wildlife. This presents an opportunity to harness digital technologies to complement conservation initiatives.

7.2.2 *Digital Technologies and Wildlife Tourism Resources*

Digital technology refers to the use of information and communication technologies and related software to capture, store, synthesise, and retrieve data. It helps in speeding up decision-making processes. Digital technologies denote a wide range of technologies, tools, services, and applications using various types of hardware and software (Tulinayo et al., 2018). The tourism sector is suited to the use of digital technologies given its service-oriented products. The present study is concerned with digital technology tools such as cameras, unmanned aerial vehicles (UAVs), drones, mobile gadgets, GPS systems, and satellites, among others in wildlife conservation space. The use of digital activism as framed by Mkono (2018) was a turning point in furthering research on African wildlife conservation. The topic is especially pertinent given that digital technology tools have become embedded in every aspect of human life; particularly so in the context of the fourth industrial revolution (4IR) (Manda & Dhaou, 2019). As the age of digital technology unfolds, understanding how sectors such as wildlife tourism resource conservation can harness and deploy the tools to enhance the conservation of wildlife resources is increasingly important.

The study leans on the technology acceptance model (TAM) to examine the extent to which digital technologies can be harnessed and deployed in the conservation of wildlife tourism resources in Zimbabwe‘s state-owned national parks. Wildlife management remains an integral component of biodiversity conservation; and as such wildlife protection is key to achieving species diversity and richness of wildlife resources (Konate et al., 2021). Digital technologies broadly include the use of personal computers, digital television, radio, mobile phones, and robots. Despite the huge benefits of digital technologies in wildlife conservation, the degree of uptake and usage is low. One of the most dominant theories in the field of digital technologies and the most widely used model for determining technology acceptance is the Technology Acceptance Model (TAM). TAM aims to determine the barriers and enablers to the adoption of new technologies in a particular domain (Tulinayo et al., 2018). The adoption of digital technologies for wildlife conservation could potentially address the need for modern conservation of wildlife. Saunders (2011) argues that the applicability and success of such technologies in effectively conserving wildlife resources varies geographically. There has been growing use of SMART technology to monitor anti-poaching patrols in many wildlife areas (Kuiper et al., 2020; Wilfred et al., 2019). However, according to a recent study by Kavhai and Mpakari (2021), the use of such technologies in developing countries such as Zimbabwe is still in its infancy. Furthermore, Wilfred et al. (2019) observe that the use of traditional conservation systems remains dominant because of a lack of capacity for emerging technologies and resources to support such technologies.

Consequently, most countries in southern Africa, including Zimbabwe, continue to grapple with serious poaching crises (Buscher & Ramutsindela, 2015; Gandiwa et al., 2014), while policing wildlife protection practices across national borders is fraught with difficulties. This has been worsened by what Lindsey et al. (2016) outlined as a massive budget deficit for conservation in Africa. However, the region's hope is now pinned on the mooted global funding model for conservation, which could be the panacea to a lack of financial resources if it materialises (Mkono, 2018).

7.3 Materials and Methods

Fieldwork was conducted at multiple sites across Zimbabwe's 10 national parks. The purpose was to examine the opportunities and challenges of harnessing digital technologies to enhance wildlife tourism resource conservation. A qualitative research approach was adopted. According to Ritchie et al. (2013), there is no single way of collecting data in qualitative research. To this end, the study is informed by an interpretive approach to the subject matter. The researchers would interpret the phenomena in terms of the meanings people assign to them (McLeod, 2019). A combination of semi-structured face-to-face and telephone interviews were conducted with key informants selected from the wildlife tourism resource conservation sector through purposive and snowball sampling procedures. Initially, a primary source contact was approached on the advice of the ZTA and Zimparks. This was augmented with the authors' extensive tourism and hospitality industry knowledge and experience to purposively identify the initial primary contact. Subsequently, the primary source suggested the next potential informant. The process was repeated over and over again. After selecting the potential interviewees, the purpose of the study was explained to the potential participants prior to the interview. Only those who provided their informed consent were then invited to participate in scheduled interviews. Appointments were made in advance. A total of 15 key informants were selected. Table 7.1 shows interviewee demographics. Where face-to-face interviews could not be secured prospective interviewees were requested to furnish their contact telephone numbers and emails to facilitate the interviews. Interview questions were informed by past studies on wildlife tourism resource conservation and research questions. A pilot study was conducted to refine the interview questions. Both telephone and face-to-face interview sessions lasted an average of 40 min. Overall, 5 interview sessions were conducted face-to-face at the workplaces of the participants (Harare, Kariba, Manapools, and Lake Chivero) while the other 10 sessions were done telephonically. Data were recorded verbatim and were later transcribed. Where interviewees expressed discomfort with being audio-recorded, data were recorded through note-taking. Interviews were continued until no new information was emerging.

To be included in the study the participant needed to have relevant working experience in wildlife tourism resource conservation and have knowledge of the relationship between digital technologies and wildlife conservation. Semi-structured

Table 7.1 Demographic data of respondents

Participant	Designation	Gender	Education	Working experience in years	Site
P1	Zimparks Area Manager	M	Masters	15 years	Chimanimani
P2	Lake Navigation Captain	M	Degree	20 years	Nyanga
P3	Park Ranger	F	HND	5 years	Kariba
P4	Ecologist	F	Masters	8 years	Mana Pools
P5	Director-Tourism Services	M	Masters	7 years	Harare
P6	Manager Non-consumptive Tourism	F	Degree	11 years	Harare
P7	I.T Manager	M	Degree	10 years	Gonarezhou National Park
P8	Network Manager	M	Degree	13 years	Matobo
P9	Game Patrol Officer	M	Diploma	8 years	Chirisa
P10	Community Resident	F	Degree	0	Chiredzi
P11	Conservation Manager	F	Masters	15 years	Hwange National Park
P12	NGO Sector Representative	F	Degree	11 years	Save Valley conservancy
P13	Digital Technology Manager	F	Degree	13 years	Victoria Falls
P14	Mapping Officer	M	HND	10 years	Lake Chivero
P15	Surveillance and Remote Sensing Officer	F	Degree	8 years	Victoria Falls

Source: Authors

interviews were preferred as they allowed researchers to ask follow-up questions. A thematic content analysis was employed to analyse data.

7.4 Presentation and Discussion of Findings

Four themes emerged from data analysis. The findings are discussed below:

7.4.1 Types of Digital Technology Tools Adopted and Deployed in Wildlife Tourism Resource Conservation

Findings showed that a wide range of digital technologies have been adopted and deployed, and are at various stages of implementation across most national parks in Zimbabwe. The digital technologies which are already in use encompass drones,

CCTV cameras, satellites, and GPS systems. The implementation started with the most problematic areas like Hwange National Park (HNP) and Gonarezhou National Park. In other national parks, pilot projects are underway, while others are being used on a trial basis. These views were summed up by the following excerpts:

> We have pilot projects in various national parks as part of the broader wildlife conservation strategy. Most digital technologies are sponsored by our developmental partners (Game Patrol Officer, P9).
>
> Drones, CCTV cameras, camera traps, satellites, metal detectors, use of night vision technologies and GPS equipment are already in use (Digital Technology Manager, P13).
>
> We have utilised some ICTs albeit at a limited scale. We need more of these digital technology tools if we are to make significant headway against poachers; especially handheld devices linked through satellite communications, cyber tracker conservation, and Radio Frequency Identification (RFID) tags (Surveillance and Remote Sensing Officer, P15).

The study established that the following specific digital technologies have already been adopted: the use of surveillance cameras to scare poachers, electronic tagging to trace and locate the animals, the use of drones to track the movement of wildlife, data storage and sharing with wildlife stakeholders such as Zimparks, NGOs, and communities. These findings are in line with the observation of the World Wildlife Fund [WWF] (2012) which predicted a general upward trend in the use of drone aircraft in tracking poachers who hunt endangered rhinoceros, elephants, and tigers. This is increasingly becoming part of the conservation strategies in most national parks. The drones are equipped with cameras which enable park rangers to locate and identify poachers (O'Donnell, 2012). Together these results provide important insights into the adoption of digital technology tools in wildlife tourism resource conservation in Zimbabwe's national parks. As threats to wildlife tourism resources evolve, countermeasures must also change. The trend in Zimbabwean national parks reflects what is prevailing in other regional countries like Kenya which has also implemented drone surveillance in many wildlife sanctuaries including the famous Maasai Mara National Reserve. This gives credence to the ideas of Spillane (2013) which suggests that drones are basically the future of conservation; a drone can do what 50 rangers can do. Spillane (2013) established that 15 elephants are now collared in the Maasai Mara reserve with a particular GPS that allows them to be tracked on a computer which overlays their paths on Google Earth. Additionally, Kenya implemented a strategy to microchip every rhino in the country. The chips are implanted into the horn of the rhino to allow tracking. Similar digital technologies were implemented in Namibia, as well as a variety of cameras and sensors that conduct daytime and nighttime reconnaissance flashes, tagging of rhinoceroses with radio frequency identification (RFID) chips, putting in place ground-based sensors connected to real-time communication, all were integrated into key areas. Kermeloitis (2013) commended the use of drones in light of their superiority over human capabilities in terms of the ability to monitor vast areas of land and therefore are ideal for national parks. Kermeloitis (2013) asserts that drones transmit a live stream to a laptop on the ground to provide information to rangers to reach hotspots quickly.

The findings are also in consonance with Chanyandura et al. (2021) who found that rhinoceros conservation strategies being implemented included education and awareness campaigns, better equipped and more anti-poaching efforts, and use of innovative systems and technologies. Technology can offer support to the natural environment, wildlife, and law enforcement agencies. Some technology is designed to be implemented on one animal at a time. Radiofrequency identification tags are microchips that are implanted into specific animals in order to track their movements through ground or mobile sensors and ensure that the animal is behaving normally. RFI tags are increasingly commonplace in rhino conservation where the chip is placed inside the rhino's horn (WWF, 2015; Edwards, 2014). Camera traps have been adopted for anti-poaching. The same applies to metal scanners aimed to detect metal traps and wire snares buried underground or hidden in vegetation thickets. Metal scanners uncover such. High-resolution images taken from satellites have been used to track illegal deforestation and protected areas encroachment in Brazil. The above shows that Zimbabwean national parks are on par with regional and international best practices in integrating ICTs and emerging digital technology tools with traditional conservation methods.

7.4.2 Challenges Faced by Wildlife Resource Conservation Authorities in Adopting Digital Technologies

A variety of challenges were outlined by respondents. Respondents conceded that wildlife authorities face numerous challenges in harnessing digital technologies, which are of technical, human resources, budgetary nature, and other context-specific challenges peculiar to communities adjacent to national parks. These militate against the full-scale adoption and deployment of digital technologies in most national parks. The responses below illustrate the diversity of challenges:

> Sabotage from local communities, funding constraints, unnavigable landscapes and terrain, and manpower shortage (NGO sector representative, P12).
>
> Vandalism of digital technology infrastructure, network and connectivity issues (Community resident, P10).
>
> Poverty among community residents, inadequate equipment needed to mount the technologies, lack of skilled personnel, exorbitant costs of necessary infrastructure, use of increasingly sophisticated weapons by poachers, lack of cooperation from communities (Manager non-consumptive tourism, P6).

From the above analysis, the traditional view that generally developing countries lack adequate resources to roll out digital technologies is reinforced (Lindsey et al., 2016). These findings are consistent with the ideas of Kavhai and Mpakari (2021), who outlined that similar challenges were encountered during the implementation of the SMART system in Mid- Zambezi particularly poor Internet connection to submit collected data to the main server for central storage. Furthermore, cyber-trackers had problems with connecting to GPS satellites when put inside pockets. Findings are also consistent with the tenets of the technology acceptance model

(Tulinayo et al., 2018) which suggest that there are factors which favour and inhibit adoption of technology. While there are huge opportunities to adopt and deploy digital technologies to successfully fight poaching, it has been observed that most African countries are more concerned about basic necessities to survive and to deal with poverty and its attendant disadvantages (Mbaiwa, 2017). This greatly affects conservation. These findings corroborate the general trend of resource scarcity in most developing countries; and conform to the issues associated with the TAM which identified a number of constraints and enablers in the acceptance of new technologies. This is consistent with Kavhai and Mpakari (2021), who found that the introduction of SMART technology was marred by technical and capacity challenges, resource challenges, and behavioural and human dynamic challenges in Mid-Zambezi.

7.4.3 Wildlife Resource Conservation Aspects Which Present Opportunities for the Adoption of Digital Technologies

Findings revealed the existence of a wide array of wildlife tourism conservation aspects, which present vast opportunities for the adoption and deployment of digital technology tools by wildlife tourism resource conservation authorities. These aspects are notably population counting, data and information sharing among the wildlife tourism conservation ecosystem, real-time access to information by wildlife tourism conservation stakeholders, quick decision-making process, snare detection, monitoring movement of mammals, mapping of water holes, data management, satellites images, and scaring away of poachers as a result of 24-h surveillance. The following responses illustrate this:

> Monitoring wildlife-ecological balance thresholds (Zimpark Manager, P1)
> Aerial wildlife population surveys- the use of drone technology to count the elephants and other faunal species with improved accuracy (Park Ranger, P3)
> Relocation of mammals in cases of overcrowding (Conservation manager, P11).

These findings overall match those observed in other previous studies. Furthermore, the findings of the present study corroborate the ideas of Matikiti-Manyevere and Rambe (2022) who suggested that the broad travel and tourism industry can benefit from virtual technologies, artificial intelligence, 5G technologies, and related technologies, particularly in times of crisis. Furthermore, the study outlines that there are early indications that Fourth Industrial Revolution (4IR) technologies have been widely deployed to address several important daily living and travel concerns. Hall et al. (2020) outline that technologies have become a major factor in building resilience in tourism. The 4IR essentially means improved production and information technologies which meet the tailored requirements of many human domains in less time, including wildlife conservation (Javaid et al., 2020). Wireless communication is provided by these technologies. The 4IR is defined by technical developments such as artificial intelligence (AI), automation, and robots as well as mobile device connectivity and data and information access. Similarly, the findings are in line with the ideas of Annecke and Masubelele (2016), UNEP (2013) and Bending (2018) who suggested that digital technologies can be deployed to augment traditional law enforcement initiatives, and innovative systems such as seismic sensors, satellites, drones, field closed circuit television cameras (CCTV), and sniffer dogs. Overall, opportunities for conservation vary across the spatial scale in a well-coordinated stakeholder and community involvement, strategic meta-population management, enhancing law enforcement initiatives through incorporating real-time surveillance technologies and intruder detection sensor networks for crime detection, and information repository for international corporations. The same pattern unfolds in the context of Zimbabwean national parks. This resonates with the prevailing circumstances in which threats to wildlife conservation efforts have been compounded by extremely limited resources. Given that most governments are likely to redirect wildlife protection resources to humanitarian considerations; Gossling et al. (2020) concede that some critical wildlife conservation programmes would inevitably face funding cuts, and would be on the brink. In view of this adoption of digital technologies becomes imperative. Carty and Onyett (2006) assert that ICTs and digital technologies ushered in new forms of communication across vast geographic distances. This resonates with wildlife conservation given the gigantic size of Zimbabwean national parks. Overall, digital technologies offer opportunities that blend the protection of wildlife resources with traditional conservation methods.

7.4.4 *Transformation of Wildlife Tourism Resource Conservation After Adopting Digital Technologies*

Of the respondents, 80% were of the belief that the adoption of digital technologies in wildlife resource conservation had significantly transformed conservation efforts as evidenced by improved efficiency and administration of resources, better data and information management and dissemination, as well as easy communication with stakeholders. The following responses illustrate this:

> Since we implemented the digital technologies we have witnessed a decrease in poaching statistics albeit at a localised level on the so-called hotspots. Overall poaching is not as widespread as it was before in already mapped sites; however, we are not sure if this is going to be sustained in the long term. (Director Tourism Services, P5).
>
> There has been marked improvement in intercepting poachers, improved coordination and collaboration among wildlife stakeholders unlike before when the stakeholders were fragmented because of communication challenges (Ecologist, P4).
>
> Ground patrol officers now have accurate data, coupled with full profiles of mammals mostly targeted by poachers. They can also cover a lot more ground in a very short space of time courtesy of digital technologies, especially sites with difficult terrain (Mapping officer, P14).
>
> Digital technologies work in issues related to deterring poaching; however, when it comes to veld fires and negative impacts of climate change, they are insignificant (Network manager, P8).

In terms of the effectiveness of digital technologies, most respondents were of the view that they are more effective compared to traditional conservation methods, especially in the aspect of detecting snares. However, respondents believed that digital technologies are not a replacement for traditional conservation methods, and emphasised that they should complement the traditional conservation methods for more effective results. Digital technologies are useful in hotspots and during the night when poachers take advantage of the cover of darkness. The finding confirms the complementarity between traditional wildlife conservation methods and the use of digital technologies. On the basis of the present study's findings, it can be stated that the adoption of digital technologies partly addresses some of the contemporary issues affecting wildlife tourism, however, it cannot be a panacea to the broader wildlife tourism resource conservation issues. For example, climate change aspects and endemic poverty levels in communities bordering wildlife resources would need broader policy reforms.

7.5 Conclusion

The study examined the opportunities and challenges of harnessing digital technologies in the conservation of wildlife tourism resources in Zimbabwean national parks. It employed a key informant interview technique to collect data from relevant wildlife tourism resource stakeholders. Overall, the study found that Zimbabwean national parks have adopted drones, CCTV cameras, satellite mapping, metal scanners, and mobile gadgets as part of the wildlife tourism resource conservation arsenal albeit at a limited scale. The digital technology tools have been implemented in most national parks with first priority given to hotspot areas. The study showed that there are vast opportunities to adopt digital technologies in a wide array of wildlife tourism resource conservation aspects. The most notable aspects were outlined as counting of mammals, detection of snares, and hard-to-reach areas, among others. Additionally, the study identified a variety of challenges faced by wildlife tourism resource conservation authorities in adopting digital technologies; which were outlined as inadequate financial resources, unskilled personnel, infrastructure vandalism, and other context-specific challenges. Findings confirmed significant transformation of the wildlife resource conservation initiatives after the adoption of digital technologies particularly in data and information management, improved collaboration among stakeholders, and apprehension of poachers. However, on the downside, the study established that digital technologies have limitations against veld fires and climate change effects. The study has contributed to the evolving wildlife resource conservation discourse by integrating the protectionist paradigm and the technology acceptance model to bring out insights around the topical aspect of wildlife resource conservation. On the basis of the study's findings, the following recommendations are given: more funding for conservation, and adoption of inclusive wildlife use approaches to address poverty issues. The study has a major limitation of using a very small sample size, which means findings lack generalisability.

Future studies may replicate the research in other geographical settings using a bigger sample size and focus on private wildlife conservancies to compare findings.

References

Adams, W. M. (2013). *Against extinction: The story of conservation*. Earth Scan.

Agu, H. U., & Gore, M. L. (2020). Women in wildlife trafficking in Africa: A synthesis of literature. *Global Ecology and Conservation, 23*, e01166. https://doi.org/10.1016/j.gecco.2020.e01166

Agyeman, Y. B., Yeboah, A. O., & Ashie, E. (2019). Protected areas and poverty reduction: The role of ecotourism livelihood in local communities in Ghana. *Community Development, 5*(1), 73–91.

Anderson, B., & Jooste, J. (2014). Wildlife poaching: Africa's surging trafficking threat. In *Africa security brief* (pp. 1–8). Africa Centre for Strategic Studies. Retrieved from ISSN2164-4039.

Annecke, W., & Masubelele, M. (2016). A review of the impact of militarisation: The case of rhino poaching in Kruger national park, South Africa. *Conservation and Society, 14*(3), 195. https://doi.org/10.4163/0972.191158

Ariya, G., Sitati, N., & Wishitemi, B. (2017). Tourists' perceived value of wildlife tourism product at Lake Nakuru national park, Kenya. *European Journal of Tourism, Hospitality and Recreation, 6*(2), 147–156.

Atkins, J. F., Maroun, W., Atkins, B., & Barura, E. (2018). From the big five to the big four? Exploring extinction accounting for the rhinoceros. *Accounting, Auditing, and Accountability Journal, 31*(2), 674–702. https://doi.org/10.1108/AAAJ-12-2015-2320

Bending, Z. J. (2018). Improving conservation outcomes: Understanding scientific, historical and cultural dimensions of the illicit trade in rhinoceros' horn. *Environment and History, 24*(2), 149–186. https://doi.org/10.3197/096734018x15137949591891

Bothma, J. D., & Du-Toit, A. J. (2016). *Game ranch management* (6th ed.). Van Schank.

Brondizio, E., Settele, J., Diaz, S., & Ngo, H. (2019). *Global assessment report on biodiversity and ecosystem services on the intergovernmental science-policy platform on biodiversity and ecosystem services*. IPBES Secretariat.

Buscher, B., & Ramutsindela, M. (2015). Green violence rhino poaching and the war to save South Africa's peace parks. *African Affairs, 115*(458), 1–22.

Carty, V., & Onyett, J. (2006). Protest, cyberactivism and new social movements: The reemergence of the peace movement post 9/11. *Social Movement Studies, 5*(3), 229–249. https://doi.org/10.1080/14742830600991586

Ceballos, S. G., Ehrlich, P. R., & Dirzo, R. (2017). Biological annihilation via ongoing sixth mass extinction signalled by vertebrate population losses and declines. *Proceedings of the National Academy of Sciences, 114*(30), E6089–E6096. https://doi.org/10.1073/pnas.1704949114

Chanyandura, A., Muposhi, V. K., Gandiwa, E., & Muboko, N. (2021). An analysis of threats, strategies and opportunities for African rhinoceros' conservation. *Ecology and Evolution*. Wiley. https://doi.org/10.1002/ece3.7536

Craigie, I. D., Baillie, J. E. M., & Balmford, A. (2010). Large mammals' population declines in Africa's protected areas. *Biological Conservation, 143*(9), 2221–2228.

Dube, K. (2021). Implications of COVID-19 induced lockdown on the South African tourism industry and prospects for recovery. *African Journal of Hospitality, Tourism and Leisure, 10*(1), 270–287.

Edwards, J. (2014). World Wildlife Fund uses RFID to Foil Poachers- A real-time tracking and monitoring solution helps protect endangered rhinos and other animals in Namibia. *Radio Frequency Identification (RFID) Journal Newsletter*.

Fynn, R., & Kolawole, O. (2020). *Poaching and the problem with conservation in Africa (commentary)*. Mongabay Environmental News.

Gandiwa, E., Zisadza-Gandiwa, P., Mango, L., & Jakarosi, J. (2014). Law enforcement stifle perceptions of illegal hunting and wildlife conservation in the Gonarezhou national park, Southeast Zimbabwe. *Tropical Ecology, 55*(1), 119–127.

Goldenberg, S. (2013). US to destroy ivory stocks in effort to stop illegal elephant poaching. *The Guardian*. Retrieved from https://www.theguradian.com/environment/2013/sep/09/us-ivory-stockspilligal-elephant-poaching

Gossling, S., Scott, D., & Hall, C. M. (2020). Pandemics, tourism and global change: A rapid assessment of Covid-19. *Journal of Sustainable Tourism, 29*(1), 1–20.

Gross, M. (2015). Can we change our predatory ways? *Comment Biology, 25*(20), R965–R967.

Haas, L. T., & Ferreira, M. S. (2018). Finding politically feasible conservation policies: The case of wildlife trafficking. *Ecological Applications, 28*(2), 473–494.

Hall, C. M., Scott, D., & Gossling, S. (2020). Pandemics, transformations and tourism: Be careful what you wish for. *Tourism Geographies, 22*(3), 577–598.

Herbig, F., & Minnaar, A. (2019). Pachyderm poaching in Africa: Interpreting emerging trends and transitions. *Crime, Law and Social Change, 71*(1), 67–82. https://doi.org/10.1007/s10611-018-9789-4

Javaid, M., Haleem, A., Vaishya, R., Bahl, S., Suman, R., & Vaish, A. (2020). Industry 4.0 technologies and their applications in fighting Covid-19 pandemic. *Diabetes and Metabolic Syndrome Clinical Research and Reviews, 14*, 419–422.

Johnson, C. N., Balmford, A., Brook, B. W., Buettel, J. C., Galettyi, M., Guang Chun, L., & Wilmshurt, J. M. (2019). Biodiversity losses and conservation responses in the Anthropocene. *Science, 356*(6335), 270–275.

Kavhai, B., & Mpakari, K. S. (2021). Spatial monitoring and reporting tool (SMART) in Mid-Zambezi Valley, Zimbabwe. Implementation challenges and practices. *Conservation Science and Practice, 3*(9), e492. https://doi.org/10.1111/csp2.492

Kermeloitis, T. (2013, January 30). *Drone ranger: Unmanned plane to spy on rhino poachers*. Retrieved from: https://www.cnn.com/2013/01/world/Africa/drone-poaching-ol-pegetal

Kishnani, N., & Sharma, V. (2022). Digitising wildlife tourism in Madhya Pradesh for sustainable value creation. *Journal of Management, 13*(1), 13–26. https://doi.org/10.18311/sdmimd/2022/29569

Konate, S., Siriman, D., Ardjima, L., Sancel, Y., Hema, E. M., Kabre, B. G., & Petrozzi, F. (2021). Longterm changes in population size and the age structure and sex ration of waterbucks in a Sudanian savannah of Burkina Faso. *African Journal of Ecology, 59*(1), 132–141. https://doi.org/10.1111/aje.12793

Kuiper, T., Kavhu, B., Ms, N. A. N., Mandisodza-Chikerema, R., & Milner-Gulland, E. J. (2020). Rangers and modellers collaboration to build and evaluate spatial models of African elephant poaching. *Biological Conservation, 243*, 108486. https://doi.org/10.1016/j.biocon.2020.108486

Lindsey, P. A., Balme, G. A., Funston, P. J., Henschel, P. H., & Hunter, L. T. (2016). Life after Cecil: Channelling global outrage into funding for conservation in Africa. *Conservation Letters, 9*, 296–301.

Lindsey, P., Allan, J., Brehony, P., Dickman, A., Robson, A., Begg, C., Bhamna, H., Blanken, L., Brener, T., & Tyrrell, P. (2021). Conserving Africa's wildlife and wildlands through Covid-19 crisis and beyond. *Nature Ecology and Evolution, 4*(10), 1300–1310. https://doi.org/10.1038/5455915

Mahoney, S. P., & Jackson, J. J. (2013). Ensuring hunting as a foundation for conservation-the North American model. *International Journal of Environmental Studies, 7*(3), 448–459.

Manda, M., & Dhaou, S. B. (2019). Responding to the challenges and opportunities in the 4th industrial revolution in developing countries. In *Proceedings of the 12th international conference on theory and practice of electronic governance. Melbourne, Australia, 3 to 5 April*.

Mareto, W. (2018). Wildlife crime: Politics, people, and prevention. *Global Environmental Politics, 19*(4), 127–131. https://doi.org/10.1162/GLEP

Matikiti-Manyevere, R., & Rambe, P. (2022). The impact of Covid-19 on the tourism industry: The role of Fourth Industrial Revolution Technologies in Mitigating the impacts. *African*

Journal of Hospitality, Tourism and Leisure, 11(3), 1173–1187. https://doi.org/10.46222/ajhtl.19770720.283

Matungwa, L. M., & Wawa, A. I. (2021). The effectiveness of anti-poaching techniques in combating wildebeest poaching in Serengeti national park. *Tanzania Journal of Forestry and Nature Conservation, 9*(3), 24–39.

Mbaiwa, J. E. (2017). Effects of the safari hunting tourism ban on rural livelihoods and wildlife conservation in Northern Botswana. *South African Geographical Journal, 100*, 41–61.

McLeod, S. A. (2019, July 30). Qualitative vs. quantitative research. *Simply Psychology*. https://www.simplypsychology.org/qualitative-quantitative.html

Mkono, M. (2018). The age of digital activism in tourism: Evaluating the legacy and limitations of the Cecil anti-trophy hunting movement. *Journal of Sustainable Tourism, 26*(9), 1608–1624. https://doi.org/10.1080/09669582.2018.1489399

Muboko, N., & Murindagomo, F. (2014). Wildlife control, access and utilisation: Lessons from legislation, policy evolution and implementation in Zimbabwe. *Journal of Nature Conservation, 22*(3), 206–211.

Muboko, N., Muposhi, V., Tarakani, T., Gandiwa, E., Vengesayi, S., & Makuwe, E. (2014). Cyanide poisoning and African elephant mortality in Hwange National Park, Zimbabwe: A preliminary assessment. *Pachyderm, 55*, 92–94.

Muboko, N., Gandiwa, E., Muposhi, V., & Tarakini, T. (2016). Illegal hunting and protected areas: Tourist perceptions on wild animal poisoning in Hwange National Park, Zimbabwe. *Tourism Management, 52*, 170–172.

Muposhi, V. N., Gandiwa, E., Bartels, P., & Makaza, S. M. (2016). Trophy hunting, conservation and rural development in Zimbabwe: Issues, options and implications. *International Journal of Biodiversity*. https://doi.org/10.1155/2016/8763980

Nellemann, C., Henriksen, R., Raxter, P., Ash, N., & Mrema, E. (Eds.). (2014). *The environmental crisis threats to sustainable development from illegal exploitation and trade in wildlife and forest resources. A UNEP rapid response assessment*. United Nations Environment Programme and GRID. Arendal, Nairobi and Arendal.

Neupane, D. (2020). How conservation will be impacted in the Covid-19 pandemic. *Wildlife Biology, 2*. https://doi.org/10.2981/wlb.00727

O'Donnell, L. (2012). *Google WWF drones: Google supplies WWF with drones to help stop poachers in Africa*. Retrieved from: http://www.policymic.com/articles/20338/google-wwf-drones-google-supplies-wwf-with-drones-to-help-stop-poachers-in-Africa

Pimms, S., Jenkins, C., Abell, R., Brooks, T., Gittleman, J., Joppa, L., Raven, P., Roberts, C., & Sexton, J. (2014). The biodiversity of species and their rates of extinction, distribution and protection. *Science, 344*, 987–997.

Quesada, B. R. (2019). Chapter 4: IPBES global assessment on biodiversity and ecosystem services. In *Plausible futures of nature, its contributions to people and their good quality of life* (pp. 1–25). IPBES.

Ramutsindela, M. (2016). Wildlife crime and state security in southern Africa: An overview of development. *Politik, 43*(2), 159–171. https://doi.org/10.1080/0259346.2016.1201376

Ritchie, J., Lewis, J., Nicholls, C. M., & Ormston, R. (Eds.). (2013). *Qualitative research practice: A guide for social science students and researchers*. Sage.

Saar, B., Sene-Harper, A., & Gonzalez-Hernandez, M. M. (2021). Tourism, social representation and empowerment of rural communities at Langue de Barbarie National park, Senegal. *Journal of Sustainable Tourism, 29*, 1383–1402.

Saunders, L. (2011). Estimating effectiveness of law enforcement on tiger prey poaching from anti-poaching patrol data. Msc Thesis in Biometry.

Schmidt, J., & Uriely, N. (2019). Tourism development and the empowerment of local communities: The case of Mitzpe Raman, a peripheral town in the Israel Negev desert. *Journal of Sustainable Tourism, 27*, 805–825.

Schwartz, M. W., Cook, C. N., Pressey, R. L., Pullin, A. S., Runge, M. C., Salafsky, N., Sutherland, W. J., & Williamson, M. A. (2018). Decision support framework and tools for conservation. *Conservation Letters, 12*(2), e12385. https://doi.org/10.1111/con/.12385
Spenceley, A., McCool, S., Newsome, D., Baez, A., et al. (2021). Tourism in protected and conserved areas amid the COVID-19 pandemic. *Parks, 27*, 103–118. https://doi.org/10.2305/IUCN
Spillane, C. (2013, October). How google earth and drones are saving elephants in Africa. *The Sydney Morning Herald*. Retrieved from: http://www.smh.com/au/technology/technology-news-how-google-earth-and-drones-are-saving-elephants-in-Africa-20131010-.2v9na.html
Stoett, P. J. (2019). *Global ecopolitica: crises, governance, and justice*. University of Toronto Press.
Travers, H., Archer, L. J., Mwedde, G., Roe, D., Baker, J., Plumptre, A. J., Rwetsiba, A., & Milner-Gulland, E. (2019). Understanding complex drivers of wildlife crime to design effective conservation interventions. *Conservation Biology, 33*(6), 1296–1306.
Tulinayo, F. P., Ssentume, P., & Najjuma, R. (2018). Digital technologies in resource constrained higher institutions of learning: A study on students' acceptance and usability. *International Journal of Educational Technology in Higher Education, 15*(36). https://doi.org/10.1186/541239-018-0117-y
UNEP. (2013). *A new eye in the sky: Eco-drones* (Global environment alert service bulletin, May 2013). United Nations Environment Programme.
Vandome, C., & Vines, A. (2018). *Tackling illegal wildlife trade in Africa: Economic incentives and approaches*. Royal Institute of International Affairs.
Waikhata, J. (2020). *The impact of Covid-19 pandemic on Africa's protected areas operations and programmes*. IUCN World Convention on Protected Areas (IUCN-WCAA).
Wilfred, P., Keyeye, H., Magigi, F. J., Kisingo, A., & Nahouyo, C. L. (2019). Challenges facing the introduction of SMART patrols in a game reserve, western Tanzania. *African Journal of Ecology, 57*, 523–530.
WWF. (2012). Living planet report 2012. Gland, WWF International. http://awsassets.panda.org/downloads/1_ipr_2012_online_full_size_single_pages_final_120516.pdf
WWF-World Wide Fund for Nature (2015). Saving forests at risk. WWF Living Forests Report, Gland, Switzerland.
Wyatt, T. (2013). *Wildlife trafficking: A deconstruction of the crime, the victims and the offenders*. Palgrave McMillian. https://doi.org/10.1057/9781137269249
Zhou, Z. (2018). The tourism sector: A bright light in Zimbabwe's depressed economic environment. *African Journal of Hospitality, Tourism and Leisure, 7*(1), 1–19.
Zhou, Z. (2019a). Trans-frontier parks as vehicles for tourism development and poverty alleviation: Lessons from southern Africa. *International Journal of Hospitality and Tourism, 12*(2), 67–81.
Zhou, Z. (2019b). Tourism progress in SADC region: Postcolonial era milestones. In M. Mkono (Ed.), *Positive tourism in Africa* (pp. 137–146). Routeledge.
Zhou, Z. (2023). Human-wildlife tourism conflict: Towards an inclusive wildlife resources utilisation approach in southern Africa. *GeoJournal, 88*, 931–949. https://doi.org/10.1007/s10708-022-10667-8

Chapter 8
Implications of Digitalisation on Value Co-creation in Restaurant Service Encounters Among Disabled People in Sub-Saharan Africa

Tawanda Makuyana, Emmanuel Ndhlovu, and Kaitano Dube

8.1 Introduction

Restaurants are a sub-sector in the hospitality business that prepares and serves food and drinks (beverages) to customers, and this has been described as McDonaldization (Symons, 2013). However, service offerings vary because of the type of target market and the business model, like fast-food, fine-dine, drive-through, and food delivery services (Symons, 2013). Restaurants are an element of tourism that fits into accessible tourism as enhanced by the role of technologies and digitalisation (Ndhlovu & Dube, 2023b; Nigg & Peters, 2022). Darcy et al. (2020) regard accessible tourism as a system that allows persons with disabilities and those without disabilities to effectively participate and get involved in tourism, including restaurants as a sub-sector. Within the same view, Egard and Hansson (2023) view digitalisation as disabling barriers experienced by persons with disabilities. Emam and Kabir (2022) added a layer of inclusive e-services from a policy perspective in Bangladesh as a means to provide a guide towards systemic implementation.

Furthermore, Cerdan and Darcy (2021) considered customer-to-customer co-creation of inclusive tourism experiences for customers with disabilities in a shared heritage context experience. Scarlett et al. (2021) proposed an additional layer that expresses the role of organisations in the technological value co-creation process of restaurant consumers. Therefore, the present study upholds Fan and Luo's (2020) characterisation of value co-creation as a service-oriented management process that identifies stakeholders for strategic and tactical integration for optimal usage of existing resources within the service systems for the benefit of involved parties (individual and enterprises) within a market exchange process for innovation.

T. Makuyana (✉) · E. Ndhlovu · K. Dube
Vaal University of Technology, Vanderbijlpark, South Africa
e-mail: kaitanod@vut.ac.za

E. Ndhlovu et al. (eds.), *Tourism and Hospitality for Sustainable Development*,
https://doi.org/10.1007/978-3-031-63069-9_8

The literature indicates that digitalisation has contributed to success and failures within a shared economy like tourism (Furunes & Mkono, 2019; Ndhlovu & Dube, 2023c). Examples of success include online ordering and door-to-door delivery services (Bonfanti et al., 2023) and process excellence (Martinez, 2019). In all the successes, Montargot (2016) believe digitalisation nurtures innovation and enhanced service encounters in the hospitality sector, especially restaurant services. Furthermore, Labus and Jelovac (2022) proposed a technology acceptance model for the restaurant business as digitalisation enhances and reinvents capacities and interfaces between restaurants and patrons. Digitalisation enables effective customer engagement and self-representation, enhancing value co-creation (Labus & Jelovac, 2022).

On the one hand, Martin et al. (2022) regard factors of digitalisation in small and medium-sized enterprises (SMEs) in this Spanish sector as (1) the personal characteristics of the entrepreneurs/managers, (2) the characteristics of the businesses, and (3) the spatial location of the restaurants. On the other hand, Martin et al. (2022) indicate that characteristics of the company, such as the number of establishments belonging to a corporate group and the employees' educational level, influence the digitalisation of restaurants. Likewise, it is observed that the digitalisation process is stimulated in inland towns compared to coastal areas in Spain.

While celebrating digitalisation and technological advancement in the restaurant industry, Ndhlovu and Dube (2023a) revealed the challenges of radical technological transitions in the restaurant industry. According to Ndhlovu and Dube (2023a), the challenges emerge from organisational culture, infrastructural barriers, education and training, and financial constraints in developing countries. It is, therefore, unclear if the contribution of digitalisation to the restaurant industry's service encounters receives similar sentiments among various sub-population groups among people with disabilities. Little research has been given to exploring the experiences of persons with disabilities during service co-production and co-consumption (encounters) in sub-Saharan Africa.

Suppose the low interest in exploring the implications of digitalisation of the restaurant industry services encounters among people with disabilities is maintained. In that case, the restaurant sub-sector nurtures unclear value co-creation in restaurant service encounters. However, the literature points to persons with disabilities as a potentially lucrative market for the tourism sector that includes restaurants (Luiza, 2010; Buhalis et al., 2023; Darcy et al., 2020; World Tourism Orgaanisation (WTO), 2020; Makuyana et al., 2020). It is, therefore, essential to have a sub-Saharan perspective to the debate of the value addition of digitalising restaurants as another dimension to accessible tourism (Liu et al., 2023). Hence, the present study aims to contribute through a scoping literature review on the implications of digitalisation on value co-creation in restaurant service encounters among disabled people in Sub-Sahara Africa between 2016 and 2023. The paper is structured to present the method used in the investigation, followed by the findings, discussion, and conclusion.

8.2 Theoretical Background

8.2.1 *Theory of Value Co-creation and Digitalisation*

Value co-creation emerged in 2004 by researchers like Prahalad and Ramaswamy. Since then, various perspectives have nurtured a debate that indicates that the value co-creation concept is evolving towards digitalisation (Luo, 2023). Therefore, value co-creation refers to the process where suppliers and consumers jointly create value within the co-production of services and products (Prahalad & Ramaswamy, 2004). Ballantyne and Varey (2006) view value co-creation as spontaneous, collaborative, and dialogical interactions, which bring together elements deemed to be separated by others to be innovative and unique and achieve competitive advantages. Payne et al. (2008) explain value co-creation as a dynamic interactive process jointly completed by suppliers and customers. Therefore, it is an exchange practice involving value-creating processes between customer and supplier interactions.

On the one hand, Füller et al. (2009) brought a virtual-based dimension by describing virtual value co-creation as an enabling activity, strengthening a person's experience of self-determination and self-efficacy. On the other hand, Gebauer et al. (2010) believe value co-creation enables customers and suppliers to define and solve problems together in the "experiential environment". Therefore, consumers actively dialogue with suppliers in the "experiential environment" to jointly create a personalised experience. Aarikka-Stenroos and Jaakkola (2012) support Gebauer et al. (2010) in that value co-creation is conceptualised as joint problem-solving involving supplier and customer integrating resources in a collaborative interaction process. Suppliers apply their specialised professional skills, methods, and judgment. At the same time, customers contribute resources, such as knowledge, to create the optimal value-in-use, that is, the best possible balance between the value-in-use to be achieved and the required sacrifices.

As time passed, Grönroos and Ravald (2011) brought a layer to value co-creation by defining it as the joint activities of stakeholders interacting directly in a binary relationship. This definition implies that direct interactions contribute to the value that emerges from one or both parties and even all parties in an extensive network. The view upholds that consumers have social support from family, friends, and relatives who form a social capital that implicates market-base enlargement.

Yi and Gong (2013) added a perspective on customer behaviour by suggesting measurement scales for value co-creation. They posited two dimensions in customer value co-creation behaviour composed of customers' information seeking, information sharing, responsible behaviour, personal interaction, and customer citizenship through giving feedback, advocacy, helping, and tolerance. In the same line of thought, Ramaswamy and Chopra (2014) postulated a value co-creation paradigm, assuming that: (1) Interaction is where value is produced. (2) Individual stakeholders create and transform value together. (3) Individuals' social resources and regional high-end hotels' network resources are leveraged at the same time. (4) An innovative participation platform is a tool to connect co-creation opportunities with joint resources. (5) New value creation capability is fostered using the capability ecosystem based on society, commerce, citizen, nature, and community networks. (6) Individual experience is the basis of value output. (7) Wealth, welfare, and happiness are the shared aspirations of the individuals who participate in co-creation. (8) Hotels must make stakeholders active creators of value.

Based on the view brought by Barrett et al. (2015), value co-creation is realised and driven by the service ecosystem, which provides a common institutional logic and structure for resource integration and service exchange of value co-creation. The service platform provides modular rules and protocols that guide the exchange of resources in value co-creation. Ranjan and Read (2016) supported the view and found two dimensions of value co-creation: co-production and value-in-use. The former includes knowledge, equity, and interaction, whereas the latter includes experience, personalisation, and relationship.

In addition, value co-creation as a process in which customers participate in value creation, which turns hostile customers into positive ones; therefore, Value co-creation reflects a culture of participation in which customers seek opportunities to contribute to their world and seek customers' insights on their services and products (Ramaswamy & Ozcan, 2018). A systems-thinking emerged from Ramaswamy and Ozcan (2018), who regard value co-creation as an interactional value-creation process within interactive system environments, including agency engagements and structuring organisations. Interactive system environments are supported by interactive platforms consisting of artefacts, persons, processes, and interfaces, and the value generated is value-in-interaction.

8.2.2 Customer Value

Research on customer value in the field of co-creation value indicates three main definitions of customer value in the existing research on co-creation value: (1) Customer value, stemming from value co-creation experience, is value-in-experience. From the aesthetic perspective, customer value co-created in the consumption field is manifested as customers' sensory pleasure, the meaning of life, and reconstructing experience (Vargo & Lusch, 2008).

(2) Customer value emerges from the interactive experience process in a socio-technical context. From the 24 phenomenological viewpoints, it can be interpreted as value-in-experience, value-in-use, and value-in-context. Kohler et al. (2011) validated this framework using virtual world scenarios. Service-dominant and service logic emphasise that value is always perceived and determined by customers based on personal experience and social context and emerges in the dynamic experience process of customers using products/services. Therefore, customer co-creation value-in-experience is defined as the value-in-use of products/services (Grönroos & Voima, 2013) or value-in-context (Edvardsson et al., 2011) in a broader sense.

(3) Customer value refers to customers' perception of expected returns. The above three viewpoints show that the ultimate purpose of value co-creation is to

create value for customers, and the essence of customer value is the unique experience that customers enjoy throughout co-creation. Therefore, following the view of Holbrook (2006), a representative scholar of customer value, this study defines customer value as an interaction-based experience with relative preferences.

8.2.3 Classification of Customer Value

Holbrook (2006) explained customer value as inclusive of economic, social, hedonic, and altruistic value categories and constructed a typology framework of customer value based on two dimensions of experience level (intrinsic or extrinsic) and experience effect object (self- or other-oriented), providing references for the research on customer co-creation value-in-experience. In addition, Nambisan and Nambisan (2008) constructed a four-dimensional experience framework in a virtual co-creation environment, namely pragmatic, empathic, sociability, and usability dimensions. In the contemporary intelligent connected environment, Novak and Hoffman's (2019) consumer experience research explored the innovative smart products as independent ontologies and classified consumer experience into self-extension, self-expansion, self-restriction, and self-reduction according to the two types of roles (agency and communion) represented by consumers and smart products. Such was a build-up on the theoretical foundation set by studies on customer co-creation value-in-experience. However, they only elaborate on the connotation or dimensions without further building a measurement model of value-inexperience co-created by customers.

Within the debate, some empirical research uses the theory of customer value/brand experience for reference and refines the dimensions of customer co-created experience/value. As a virtual brand community is the best context for customers to interact with enterprises and with each other and then create value together, scholars proposed, after using the classic framework of brand experience for reference, that customer co-creation experience based on the virtual brand community includes five dimensions: sensory experience, emotional experience, thinking experience, behavioural experience, and relational experience. Based on the early two-dimension typology of customer value, utilitarian value, and hedonic value (Holbrook & Hirschman, 1982), some scholars summarised value-in-experience co-created by customers as functional/practical value and hedonic/emotional value and extended it to three dimension and four-dimension values.

Other empirical research constructs the dimensions of customer co-creation value-in-experience based on a specific co-creation context. For a virtual customer environment, Nambisan and Baron (2007) operationalised customers' perceived value in participating in product support activities into learning benefits, social integrative benefits, personal integrative benefits, and hedonic benefits. On this basis, Verleye (2015), from the perspective of customer participation in new product/service development, classified customer co-creation experience into four dimensions: cognitive, social/personal, hedonic, and pragmatic/economic.

In recent years, technologies like artificial intelligence, big data, and the Internet of Things (IoT) have significantly changed the value co-creation model. Studies have examined customer co-creation experience/value based on different technological scenarios, such as new retail in the IoT environment, tourism platforms, and intelligent service scenarios (Balaji & Roy, 2017). Holbrook's typology framework is adopted because it applies to most dimensions. There are different approaches to defining customer value (Carlson et al., 2018; Chan et al., 2010; Dennis et al., 2017; Laud & Karpen, 2017), but the framework of Holbrook (2006) can provide a useful, practical taxonomy to classify all existent definitions (Baumann & Le Meunier-FitzHugh, 2015).

Customer value is defined as a preferred experience from some interaction and has two dimensions, i.e., extrinsic versus intrinsic value and self-oriented versus other-oriented value (Holbrook, 2006). The combination of these two dimensions can generate four types of customer values (Holbrook, 2006): (i) Economic value, which is both external and self-oriented, is embodied when products or services act as a tool to help customers to achieve their goals (such as high efficiency and high quality); (ii) Hedonic value, which is both internal and self-oriented, is an intrinsic value deriving from consumption experience and is mainly reflected in deriving pleasure from various entertainment activities, or aesthetic enjoyment from art, entertainment activities, and scenery; (iii) Social value, which is both external and other-oriented, is represented when a product or service is used as a means to influence others' reactions. It stems from two purposeful situations: one is good impression consumption to promote one's identity and social position, and the other is to gain respect by taking possession of material goods; (iv) Altruistic value, which is both internal and other-oriented, not only requires customers to regard shopping experience as the ultimate consumption goal but also needs customers to care about the influences of their consumptive behaviours on others. What should be noted is that only economic, hedonic, and social values are included in the primary studies.

8.2.4 Dimensions in Value Co-creation and Disability

The chapter acknowledges value co-creation dimensions as having three main characteristics: (1) Researchers still need to form a unified categorisation of value co-creation dimensions. (2) "Interaction", "participation", and "behaviour" are the three core dimensions highlighted by most research. (3) Value co-creation dimensions are proposed in different contexts; however, Prahalad and Ramaswamy's (2004) Dialog, Access, Risk Assessment, and Transparency (DART) model is widely used. In addition, DART is regarded as the core dimension of value co-creation, while other dimensions have specific manifestations in different contexts.

Therefore, the above section reflected that the definitions share similarities in that they all show that value co-creation is a "process" that has direct and indirect interaction to determine how to satisfy a customer. Within the hospitality context, the definition of value co-creation in academic circles has evolved along two paths.

First, "binary relationship → multiple-party relationship → network relationship in the service ecosystem" (Luo, 2023). The evolution relates to how Darcy et al. (2020) reflected the evolution of disability tourism from being a niche to mainstream as accessible tourism. In tourism and hospitality- the food and beverage function contributes to guest experiences from a restaurant's co-production and co-consumption processes (value-co creation). The value co-creation concept that is now anchored in network relationships speaks to the positive of including access-need tourism market like ageing persons with disabilities because they do not travel alone, and when they feel included, they tend to have more extended stays (Makuyana et al., 2020; WTO, 2020).

The second is "traditional value co-creation → digital value co-creation → interactive system co-creation". It, therefore, acknowledges that the environment involved in understanding value co-creation changes from the initial "offline environment" to the "digital environment" in the network era to the "interactive system" integrating the offline and digital environments (Luo, 2023). In this digital transformation space, African tourism and hospitality are still at the infant stage in adopted socio-economic transformation that seeks to mainstream the previously marginalised population groups (Makuyana & Nzo, 2022). Nonetheless, the debate on value co-creation in the tourism and hospitality spheres is still dominated by hotels based in Europe, America, and Oceania (Australia and Asia) (Luo, 2023). Restaurants have not been explored within the sector, especially from a disability inclusion (accessible tourism lens).

The International Classification of Functioning, Disability, and Health- World Health Organization (WHO) interprets disability as the adverse interactive relationship between a person's health condition(s) and contextual environmental and personal factors (WHO, 2001). The United Nations (2006) (UN) further clarifies that persons with disabilities should include those with long-term mental, physical, sensory, or intellectual impairments that may hinder their equal, full, and effective social participation when interacting with others and the environment. People with disabilities, for example, experience more challenges when travelling alone, especially when navigating crowds and ordering food in restaurants without braille menus (Loi & Kong, 2016; Magistretti, 2019; Roger, 2019).

8.2.5 Digitalisation of Restaurants and Persons with Disabilities

Value co-creation is a borrowed concept that also applies in the hospitality sector: (1) Clarify the process and structure of value co-creation (Luo, 2023). Navarro et al. (2014) incorporated the value co-creation process of customer groups. They identified factors and dynamic processes in the interaction between disabled customers and hotel services through service-dominant logic thinking. Based on the value co-creation process model and the analytic hierarchy process method, Navarro et al.

(2015) further found that the success factors of value co-creation by hotels and disabled customers are related to different stages of value co-creation. The key to successful value co-creation lies in the capacity of frontline staff to identify customer demand (knowledge of customer demand). They used qualitative and quantitative research methods to investigate the hotels' customers and frontline service staff. They explored the antecedents of the knowledge of frontline hotel staff on customer demand.

(2) Explore the impact of technology and equipment on value co-creation in the hotel industry. Based on the sample data of American hotel 22 customers, Morosan and DeFranco (2016) developed and validated a conceptual model in which hotels use mobile devices to co-create value with customers. The model recognises the central role of the degree of value co-creation by consumers when hotels try to create valuable service experiences by using mobile devices. Morosan and DeFranco (2019) researched hotel interactive technologies for value co-creation. They found that when a hotel adopts interactive technologies, consumers' information system habits and hedonic motivation will affect customer participation in value co-creation. Then, customer participation and innovation in value co-creation will impact conversion behaviour, and innovation and perceived benefits of hotel interactive technologies will influence the customers' intentions to use interactive technologies. Based on the social information processing theory, the personification of service robots in the hotel industry influences customers' intentions in value co-creation (Morosan & DeFranco, 2019).

(3) Examine the results of value co-creation in the hotel industry. Harkison (2018) interviewed 81 participants from six New Zealand luxury hotels in a case study and found that managers, employees, and customers agreed that the luxury accommodation experience is co-created by managers, employees, and customers and that encouraging co-creation can facilitate the generation of luxury accommodation experiences. Santos-Vijande et al. (2018) discussed the influence of new service co-creation on the performance of new services in the hotel industry. The research findings showed that co-creation will directly affect the market results and development speed of new services and thus affect the quality of new services. Based on 252 questionnaires, they verified the coordination complexity brought by customer 23 participation in value co-creation in developing new services in the hotel industry. They concluded that the hotel's internal and external services' heterogeneity will intensify or weaken customer participation's impact. Liu and Jo (2020) constructed a conceptual model of value co-creation behaviour, member satisfaction, customer involvement, and engagement. Their empirical survey of hotels in North America showed that value co-creation behaviour plays a significant role in influencing customer satisfaction. González-Mansilla et al. (2019) explored the management process of value co-creation in a hotel. They noted that the degree of customer participation largely depends on their perception of how the hotel supports the value co-creation process. The perception will yield a direct positive impact on the brand equity of the hotel and also on the perceived value through customer participation. Meanwhile, brand equity and perceived value are positively correlated with customer satisfaction.

8.3 Materials and Method

A scoping review that embeds the theory of value co-creation principles guides the discourse to unpack perspectives within extant literature. The researchers conducted an electronic search in internationally focused databases Google Scholar after finding no literature in SCOPUS, with Sub-Saharan research on the topic under study. An inclusion and exclusion criteria, including Sub-Saharan focused material on digitalisation and inclusion of access-need market (diverse disabled people).

Selected data will be analysed using Creswell's qualitative analysis framework (Fig. 8.1).

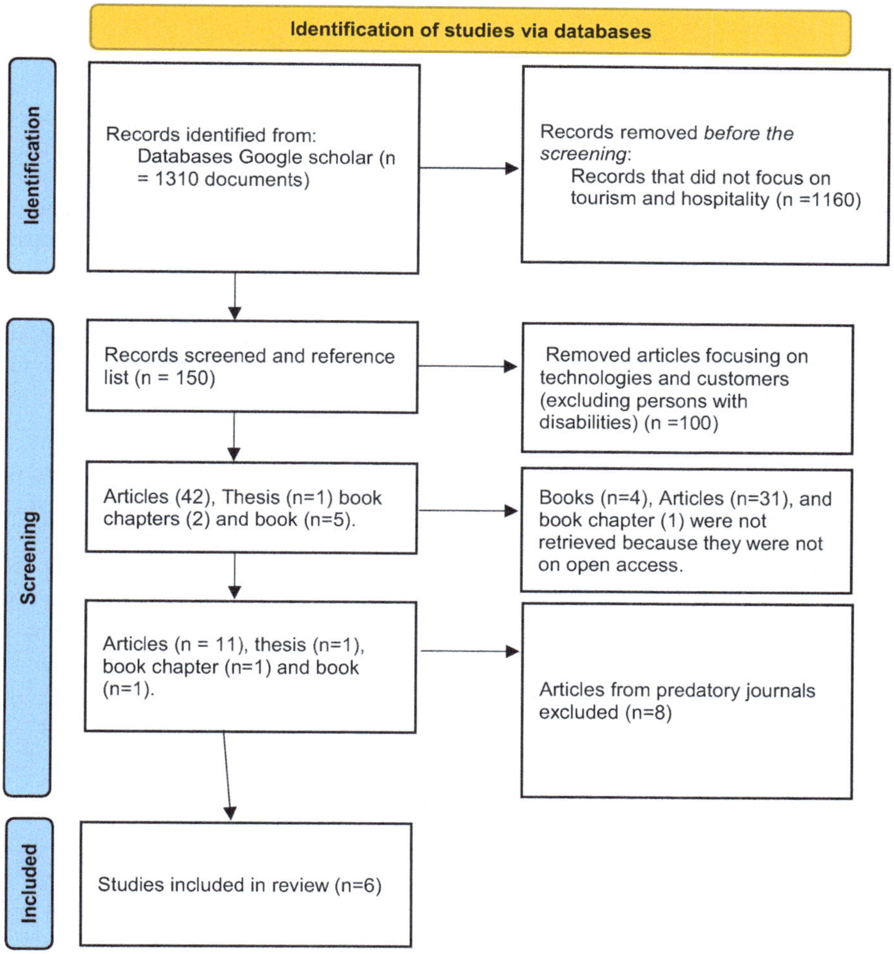

Fig. 8.1 Qualitative analysis framework. (Source: Adopted in parts from Page et al., 2020)

8.4 Findings and Discussion

The findings reflect four main themes discussed as follows:

(i) Increase research on accessible tourism: International organisation reports and academic research indicating potential value from engaging customers with disabilities. This number steadily increases, representing an enormous business opportunity for global industries (Purple, 2022).
(ii) The digitalisation of restaurant service is still to consider all forms of disabilities: there is growth in disability and technologies in hospitality, but disability research that focuses on digitalised restaurant service encounters (experiences) is still scarce. Tourism as one of the major economic sectors on the planet (WTTC, 2019) still frequently neglects (Darcy, 2010; Scheyvens & Biddulph, 2018) the emerging demand for accessible tourism (Sisto et al., 2021) and people with disabilities continue to be excluded from travel (Nazari Orakani et al., 2021).
(iii) Disability inclusion in the sub-Sahara narrative on digitalised restaurant services: The narrative that is reflected in extant literature reflects the Asian, European, and American context, leaving the sub-Saharan African narrative to have an absence of scientifically documented baseline research in this regard.
(iv) Value co-creation and disability-inclusive digitalised restaurant services: research reflects anecdotical evidence that can motivate digitalised restaurant organisations to explore access-need restaurant markets' experience. Despite that more than 15% of the global population has some form of disability (World Health Organization, 2020), the digitalised restaurant sub-sector is still to properly interpret disability within a business lens.

WTO (2020) considered accessible tourism as one of the game changers in tourism recovery during the post-health pandemic and lockdowns. The United Nations Department of Economics and Social Affairs Disability (2022) calls for the development of accessible tourism, providing equal opportunities for people with disabilities to travel. For instance, 2.2 billion people are visually impaired, and up to 45 million are blind (World Health Organization, 2021a, b). China has the largest visually impaired and blind populations at 274.3 million and 9.2 million, respectively, and most sub-Saharan states, like South Africa, are members of Brazil, Russia, India, China, and South Africa's geopolitical-economic group; thereby, if developing states such as Zimbabwe in sub-Sahara are like China, with limited accessible infrastructure, many persons with disabilities such as blind people struggle to be recognised (Yang, 2021).

Despite the increased research on disability, the digitalisation of the restaurant sub-sector has blanks in scientific documentation of encounters of persons with disabilities (customers) in sub-Saharan. Such baseline knowledge can add a layer to the disability, digitalisation, and hospitality-restaurant narratives contributed by Asian, European, and American-based research. Thus, for example, Kalargyrou et al. (2018) found that customers with family members or friends with disabilities showed less stereotyping than customers who did not. Kim and Lehto (2012) investigated disabled travellers' service failure experiences using the critical incident technique. They found that poor design for handicapped access, unaccommodated special requests, and negative behaviours of employees are among the top reasons causing service failures in the restaurant, hospitality, and tourism service sectors.

Technologies make things possible, and the value co-creation theory offers a new perspective for examining the restaurant service experiences of people with disabilities. However, despite research on value co-creation in hotels (Luo, 2023), few applied the value co-creation digitalisation of restaurant service delivery from all groups of persons with disabilities' perspectives. By focusing on restaurant service encounters among disabled people, the study explored how digital-based technologies add value to all groups of customers with disabilities' service experiences.

On the other hand, technological advancements have made people with disabilities more empowered than ever. For instance, voice-activated chatbots allow visually impaired people to control different devices used at home. Advanced GPS tools, such as RightHear, provide multilingual audio descriptions of indoor and outdoor environments. Therefore, it is critical and timely to examine how technologies can enhance the accessibility of all customers with disabilities to use restaurant services while enabling and enhancing their experiences. With the introduction of service robots and related technologies, Tlili et al. (2021) suggested integrating technology into service to accommodate people with disabilities. However, based on various disabilities, service encounters vary among each disability group (Liu et al., 2023). Research on service encounters within the digitalised restaurant industry has not covered all disabilities, both visible and invisible (Yang, 2021).

The chapter establishes a research agenda for the research community to explore. At the same time, the chapter unpacks insights for managers to review their management and operational practices and policies from an accessible tourism and hospitality lens. Such can leverage digitalisation that nurtures positive service encounters that, in turn, enable repeat business from both disabled people (access-need markets) and their support structures.

8.5 Conclusion

This chapter concludes that digitalisation has mixed value co-creation in restaurant service encounters among various disabled people in Sub-Sahara Africa due to the diversity of impairments and their digital support/needs during service encounters, attitudes, and level of digital inclusion. It implies that digitalisation does not uphold disability inclusion unless it embeds a universal design with the context of the type of restaurants and their access-need target market segments. It is, therefore, critical to bringing socio-economic context and level of digitalisation in the restaurant industry and their offerings in sub-Saharan Africa. Such has implications in the marketing and communication that includes persons with disabilities, the environmental setting, know-how, and attitude among the restaurant staff and owners' (investors) readiness to include access-need markets. It is critical to conceptualise the digitalisation of restaurants within the external dimensions, personal dimensions, and technology integration.

Thus, business owners provide the external dimensions that include restaurants with service robots and give customers to interact with when they stay at the hotel (Zhang et al., 2022). Personal dimensions refer to restaurant mobile apps as supportive technologies that customers can obtain and operate personally using smartphones (Ahn & Kwon, 2021). In sub-Saharan Africa, apps like Mr. D and restaurant chains/groups have apps used via customers' devices to make food delivery orders. Technology integration with supportive technologies requires connections between the technology devices of a business context and the technology devices of customers. For example, a digital room key is a technology integration that requires connecting customers' mobile phones and the digital sensor of a restaurant or hotel room door (Lei et al., 2019).

The limitations of the chapter are centred on having a blanket approach; however, future research should disaggregate all sub-groups within a population with disabilities, age, gender, geospatial distribution, and geopolitics of disabilities within the tourism, travel, hospitality, and events industry.

References

Aarikka-Stenroos, L., & Jaakkola, E. (2012). Value co-creation in knowledge intensive business services: A dyadic perspective on the joint problem solving process. *Industrial Marketing Management, 41*(1), 15–26.

Ahn, J., & Kwon, J. (2021). Examining the relative influence of multidimensional customer service relationships in the food delivery application context. *International Journal of Contemporary Hospitality Management, 33*(3), 912–928.

Balaji, M. S., & Roy, S. K. (2017). Value co-creation with Internet of things technology in the retail industry. *Journal of Marketing Management, 33*(1-2), 7–31.

Ballantyne, D., & Varey, R. J. (2006). Creating value-in-use through marketing interaction: the exchange logic of relating, communicating and knowing. *Marketing theory, 6*(3), 335–348.

Barrett, M., Davidson, E., Prabhu, J., & Vargo, S. L. (2015). Service innovation in the digital age. *MIS Quarterly, 39*(1), 135–154.

Baumann, J., & Le Meunier-FitzHugh, K. (2015). Making value co-creation a reality–exploring the co-creative value processes in customer–salesperson interaction. *Journal of Marketing Management, 31*(3-4), 289–316.

Buhalis, D., Lin, M. S., & Leung, D. (2023). Metaverse as a driver for customer experience and value co-creation: Implications for hospitality and tourism management and marketing. *International Journal of Contemporary Hospitality Management, 35*(2), 701–716.

Bonfanti, R. C., Tommasi, F., Ceschi, A., Sartori, R., & Ruggieri, S. (2023). The antecedents of the technology acceptance model in microentrepreneurs' intention to use social networking sites. *European Journal of Investigation in Health, Psychology and Education, 13*(7), 1306–1317.

Carlson, J., Rahman, M., Voola, R., & De Vries, N. (2018). Customer engagement behaviours in social media: capturing innovation opportunities. *Journal of Services Marketing, 32*(1), 83–94.

Cerdan Chiscano, M., & Darcy, S. (2021). C2C co-creation of inclusive tourism experiences for customers with disability in a shared heritage context experience. *Current Issues in Tourism, 24*(21), 3072–3089.

Chan, F. K., Thong, J. Y., Venkatesh, V., Brown, S. A., Hu, P. J., & Tam, K. Y. (2010). Modeling citizen satisfaction with mandatory adoption of an e-government technology. *Journal of the Association for Information Systems, 11*(10), 519–549.

Darcy, S. (2010). Inherent complexity: Disability, accessible tourism, and accommodation information preferences. *Tourism Management, 31*(6), 816–826. https://doi.org/10.1016/j.tourman.2009.08.010

Darcy, S., McKercher, B., & Schweinsberg, S. (2020). From tourism and disability to accessible tourism: a perspective article. *Tourism Review., 75*(1), 140–144.

Edvardsson, B., Tronvoll, B., & Gruber, T. (2011). Expanding understanding of service exchange and value co-creation: a social construction approach. *Journal of the Academy of Marketing Science, 39*, 327–339.

Egard, H., & Hansson, K. (2023). The digital society comes sneaking in. An emerging field and its disabling barriers. *Disability & Society, 38*(5), 761–775.

Emam, J., & Kabir, D. M. H. (2022) Inclusive e-Service or Risk of Digital Divide The Case of National ICT Policy 2018 of Bangladesh. *International Journal of Innovative Science and Research Technology (IJISRT), 7*(9), 20–30.

Fan, X., & Luo, Y. (2020). Value co-creation: A literature review. *Open Journal of Social Sciences, 8*(2), 89–98.

Füller, J., Mühlbacher, H., Matzler, K., & Jawecki, G. (2009). Consumer empowerment through internet-based co-creation. *Journal of Management Information Systems, 26*(3), 71–102.

Furunes, T., & Mkono, M. (2019). Service-delivery success and failure under the sharing economy. *International Journal of Contemporary Hospitality Management, 31*(8), 3352–3370.

Gebauer, H., Johnson, M., & Enquist, B. (2010). Value co-creation as a determinant of success in public transport services: A study of the Swiss Federal Railway operator (SBB). *Managing Service Quality: An International Journal, 20*(6), 511–530.

Grönroos, C., & Ravald, A. (2011). Service as business logic: implications for value creation and marketing. *Journal of Service Management, 22*(1), 5–22.

Grönroos, C., & Voima, P. (2013). Critical service logic: making sense of value creation and co-creation. *Journal of the Academy of Marketing Science, 41*, 133–150.

González-Mansilla, Ó., Berenguer-Contrí, G., & Serra-Cantallops, A. (2019). The impact of value co-creation on hotel brand equity and customer satisfaction. *Tourism Management, 75*, 51–65.

Harkison, T. (2018). The use of co-creation within the luxury accommodation experience–myth or reality?. *International Journal of Hospitality Management, 71*, 11–18.

Holbrook, M. B. (2006). Consumption experience, customer value, and subjective personal introspection: An illustrative photographic essay. *Journal of Business Research, 59*(6), 714–725.

Holbrook, M. B., & Hirschman, E. C. (1982). The experiential aspects of consumption: Consumer fantasies, feelings, and fun. *Journal of Consumer Research, 9*(2), 132–140.

Kalargyrou, V., Barber, N. A., & Kuo, P.-J. (2018). The impact of disability on guests' perceptions of service quality delivery in the hospitality industry. *International Journal of Contemporary Hospitality Management, 30*(12), 3632–3655.

Kohler, T., Fueller, J., Matzler, K., Stieger, D., & Füller, J. (2011). Co-creation in virtual worlds: The design of the user experience. *MIS Quarterly*, 773–788.

Kim, S. E., & Lehto, X. Y. (2012). The voice of tourists with mobility disabilities: Insights from online customer complaint websites. *International Journal of Contemporary Hospitality Management, 24*(3), 451–476.

Labus, P., & Jelovac, D. (2022). Restaurants: Applying an extended technology acceptance model. *Acta Turistica, 34*(1), 51–82.

Laud, G., & Karpen, I. O. (2017). Value co-creation behaviour–role of embeddedness and outcome considerations. *Journal of Service Theory and Practice, 27*(4), 778–807.

Lei, S. I., Wang, D., & Law, R. (2019). Hoteliers' service design for mobile-based value cocreation. *International Journal of Contemporary Hospitality Management, 31*(11), 4338–4356.

Liu, A., Ma, E., Wang, Y. C., Xu, S., & Grillo, T. (2023). AI and supportive technology experiences of customers with visual impairments in hotel, restaurant, and travel contexts.

International Journal of Contemporary Hospitality Management. https://doi.org/10.1108/IJCHM-10-2022-1243

Loi, K. I., & Kong, W. H. (2016). Tourism for all: Challenges and issues faced by people with vision impairment. *Tourism Planning and Development, 14*(2), 181–197.

Luiza, S. M. (2010). Accessible tourism–the ignored opportunity. *Annals of Faculty of Economics, 1*(2), 1154–1157.

Luo, P. (2023). *Study on visibility promotion for regional high-end hotels based on the value co-creation theory* (pp. 1–149). Dissertations and Theses Collection (Open Access). Available at: https://ink.library.smu.edu.sg/etd_coll/475

Magistretti, B. (2019). *Travelling with a visual impairment: Can we make accessibility trendy?* Available at: https://suitcasemag.com/articles/travelling-visual-impairment-can-accessibilitytrendy

Makuyana, T., Du Plessis, E., & Chikuta, O. (2020). Profiling disabled people as unrealised potentially lucrative tourism market segment in South Africa. *African Journal of Hospitality, Tourism and Leisure, 9*(5), 876–891.

Makuyana, T., & Zinhle Nzo, Z. Y. (2022). Inclusive community development through events: Lessons from casual day with persons with disabilities event in South Africa. In *Festival and Event Tourism: Building Resilience and Promoting Sustainability* (pp. 85–101). GB: CABI.

Martinez, C. (2019). Promoting critical digital literacy in the leisure-time center: Views and practices among Swedish leisure-time teachers. *Nordic Journal of Digital Literacy, 14*(3-4), 134–146.

Martín Martín, D., Maya Garcia, J., & Romero Luna, I. (2022). Determinants of digital transformation in the restaurant industry. *Amfiteatru Economic, 24*(60), 430–446.

Morosan, C., & DeFranco, A. (2019). Co-creation of value using hotel interactive technologies: examining intentions and conversion. *International Journal of Contemporary Hospitality Management, 31*(3), 1183–1204.

Montargot, N. (2016). Digitalisation Advances and Hospitality Service Encounters. *Symphonya. Emerging Issues in Management, 1*, 64–74.

Nambisan, S., & Baron, R. A. (2007). Interactions in virtual customer environments: Implications for product support and customer relationship management. *Journal of Interactive Marketing, 21*(2), 42–62.

Nambisan, S., & Nambisan, P. (2008). How to profit from a better virtual customer environment. *MIT Sloan Management Review*.

Navarro, S., Andreu, L., & Cervera, A. (2014). Value co-creation among hotels and disabled customers: An exploratory study. *Journal of Business Research, 67*(5), 813–818. https://doi.org/10.1016/j.jbusres.2013.11.050

Navarro, S., Garzón, D., & Roig-Tierno, N. (2015). Co-creation in Hotel-disable customer interactions. *Journal of Business Research, 68*(7), 1630–1634. https://doi.org/10.1016/j.jbusres.2015.02.007

Nazari Orakani, S., Smith, K., & Weaver, A. (2021). Reframing the experiences of travellers with mobility impairments: Enhancing the leisure constraints model. *Journal of Hospitality and Tourism Management, 47*, 84–92. https://doi.org/10.1016/j.jhtm.2021.03.003

Ndhlovu, E., & Dube, K. (2023a). Challenges of radical technological transition in the restaurant industry within developing countries. *African Journal of Hospitality, Tourism and Leisure, 12*(1), 156–170.

Ndhlovu, E., & Dube, K. (2023b). Impact of COVID-19 on tourism in South Africa: Building resilience and prospects for recovery. In K. Dube, G. Nhamo, & M. P. Swart (Eds.), *COVID-19, tourist destinations and prospects for recovery volume one: A global perspective* (pp. 81–100). Springer Nature: Switzerland.

Ndhlovu, E., & Dube, K. (2023c). Restaurants industry recovery challenges from COVID-19 Amidst the Russo-Ukrainian War. In K. Dube, G. Nhamo, & M. P. Swart (Eds.), *COVID-19, Tourist destinations and prospects for recovery volume one: A global perspective* (pp. 337–356). Springer Nature: Switzerland.

Novak, T. P., & Hoffman, D. L. (2019). Relationship journeys in the internet of things: a new framework for understanding interactions between consumers and smart objects. *Journal of the Academy of Marketing Science, 47*, 216–237.
Nigg, J. J., & Peters, M. (2022). The evolution of ICTs in accessible tourism: A stakeholder collaboration analysis. *Journal of Hospitality and Tourism Management, 52*, 287–294.
Page, M. J., McKenzie, J. E., Bossuyt, P. M., Boutron, I., Hoffmann, T., Mulrow, C. D., Shamseer, L., & Moher, D. (2020). Mapping of reporting guidance for systematic reviews and meta-analyses generated a comprehensive item bank for future reporting guidelines. *Journal of Clinical Epidemiology., 118*, 60–68.
Payne, A. F., Storbacka, K., & Frow, P. (2008). Managing the co-creation of value. *Journal of the Academy of Marketing Science, 36*, 83–96.
Prahalad, C. K., & Ramaswamy, V. (2004). *The future of competition: Co-creating unique value with customers*. Harvard Business Press.
Purple. (2022). *The purple pound: Facts and figures*. Purple Zest Limited. https://wearepurple.org.uk/the-purple-pound-infographic/
Ramaswamy, V., & Chopra, N. (2014). Building a culture of co-creation at Mahindra. *Strategy & Leadership, 42*(2), 12–18.
Ramaswamy, V., & Ozcan, K. (2018). What is co-creation? An interactional creation framework and its implications for value creation. *Journal of Business Research, 84*, 196–205.
Ranjan, K. R., & Read, S. (2016). Value co-creation: concept and measurement. *Journal of the Academy of Marketing Science, 44*, 290–315.
Santos-Vijande, M. L., López-Sánchez, J. Á., & Pascual-Fernandez, P. (2018). Co-creation with clients of hotel services: the moderating role of top management support. *Current Issues in Tourism, 21*(3), 301–327.
Scarlett, H. G. (2021). Tourism recovery and the economic impact: A panel assessment. *Research in Globalization, 3*, 100044.
Scheyvens, R., & Biddulph, R. (2018). Inclusive tourism development. *Tourism Geographies, 20*(4), 589–609. https://doi.org/10.1080/14616688.2017.1381985
Sisto, R., Cappelletti, G. M., Bianchi, P., & Sica, E. (2021). Sustainable and accessible tourism in natural areas: A participatory approach. *Current Issues in Tourism*, 1–18. https://doi.org/10.1080/13683500.2021.1920002
Symons, M. (2013). The rise of the restaurant and the fate of hospitality. *International Journal of Contemporary Hospitality Management, 25*(2), 247–263.
Tlili, A., Altinay, F., Altinay, Z., & Zhang, Y. (2021). Envisioning the future of technology integration for accessible hospitality and tourism. *International Journal of Contemporary Hospitality Management, 33*(12), 4460–4482.
United Nations. (2006). *United Nations enable. Draft convention on the rights of persons with disabilities*. Available at: www.un.org/esa/socdev/enable/rights/ahc8adart.htm#art2
United Nations Department of Economics and Social Affairs Disability. (2022). *Promoting accessible tourism for all*. Available at: www.un.org/development/desa/disabilities/issues/promotingaccessible-tourism-for-all.html
Vargo, S. L., & Lusch, R. F. (2008). Service-dominant logic: continuing the evolution. *Journal of the Academy of Marketing Science, 36*, 1–10.
Verleye, K. (2015). The co-creation experience from the customer perspective: its measurement and determinants. *Journal of Service Management, 26*(2), 321–342.
World Health Organization. (2001). *International Classification of Functioning, Disability, and Health*. World Health Organization.
World Health Organization. (2020). *Disability*. https://www.who.int/topics/disabilities/en/
World Health Organization. (2021a). *Blindness and vision impairment*. Available at: www.who.int/news-room/fact-sheets/detail/blindness-and-visual-impairment
World Health Organization. (2021b). *Disability and health*. https://www.who.int/en/news-room/fact-sheets/detail/disability-and-health
World Tourism Organization. (2020). *UNWTO Inclusive Recovery Guide – Sociocultural Impacts of Covid-19, Issue I: Persons with Disabilities*. UNWTO. https://doi.org/10.18111/9789284422296

WTTC. (2019). *Economic impact reports*. https://wttc.org/Research/Economic-Impact

Yang, T. (2021). *Hidden in plain sight: China's visually impaired still struggle to be seen. The world of Chinese*. Available at: www.theworldofchinese.com/2021/12/hidden-in-plain-sight-chinasvisually-impaired-still-struggle-to-be-seen/

Yi, Y., & Gong, T. (2013). Customer value co-creation behavior: Scale development and validation. *Journal of Business Research, 66*(9), 1279–1284.

Zhang, X., Balaji, M. S., & Jiang, Y. (2022). Robots at your service: Value facilitation and value cocreation in restaurants. *International Journal of Contemporary Hospitality Management, 34*(5), 2004–2025.

Chapter 9
The Battle of Hotel Accommodation Booking Technologies: Challenges of OTA's Versus Direct Bookings Systems—A Manager's Perspective

Tshinakaho Nyathela-Sunday, Ilhaam Banoobhai-Anwar, and Desré Draper

9.1 Introduction

The tourism industry is recognised as information-based and intensive and, as such, relies on Internet Communication Technology (ICT). Therefore, the Internet has been the global source of information for tourism, which plays a vital role during the booking, planning, and execution phases of a guest cycle. In the past, before the prevalence of information technology and its subsequent disruption (Merkert & Hakim, 2022) on booking systems, hotels had the infrastructure and the means to capture their target market. Preceding the development of OTA's, there were limited avenues for a guest to book a room for accommodation, namely: Walk-in, Direct—via telephone or e-mail or through a travel agent which had a vis-à-vis relationship with a hotel group (Liu et al., 2023). The vitality of ICT in tourism showcases the intangible products and services offered as it ensures accurate information (Anwar et al., 2014). It impacts marketing and distribution, specifically within the hospitality industry, involving presenting services and products, price, promotion, selling, and direct customer engagement (Schegg et al., 2013).

Although the Internet is vital for tourism, the introduction of OTAs changed the distribution and price of hotel products (Gazzoli et al., 2008). Hotel management has been blindsided by the monopolisation of foreign-owned websites or OTA's in the hotel booking industry (Anwar et al., 2014), causing a lot of financial challenges due to price differences and inconsistency (Gazzoli et al., 2008). Hotels with a higher grading, namely four-and-five-star hotels, are among the first things guests consider when booking (Tefera & Govender, 2015). This study focuses on four-and-five-star hotels graded by the Tourism Grading Council in South Africa (TGCSA),

T. Nyathela-Sunday (✉) · I. Banoobhai-Anwar · D. Draper
Cape Town Hotel School, Cape Peninsula University of Technology, Cape Town, South Africa
e-mail: NYATHELAT@cput.ac.za

E. Ndhlovu et al. (eds.), *Tourism and Hospitality for Sustainable Development*,
https://doi.org/10.1007/978-3-031-63069-9_9

which established its internationally accredited quality assurance body in 2000 that assesses various establishments against set criteria and awards star ratings accordingly (Tomlinson et al., 2023).

Direct bookings allow hotels to retain ownership of their user data (Shariffuddin et al., 2023) and have loyal customers and travel agents whom they work with who receive discounts and other promotions, resulting in return bookings and return customers through an already vetted channel (Heuvel 2022). Callan's study (1995), conducted amongst various hotel guests, found that leisure users were shown to book directly to ensure they get the best amenities possible, especially at four-and-five-star hotels, which often add value if you book directly.

Direct booking allows hotels to control credibility, an essential factor to consider. South Africa has seen a few cases involving Instagram-based travel agents allegedly disappearing without delivering on exotic overseas trips booked and paid for in full. The cost of these trips was in the millions of rands, and the money has yet to be recovered. These indirect booking channels were heavily promoted online by local influencers and celebrities and are another argument for why direct bookings should be announced as the preferred method of booking channels (Naidoo, 2022; Rall, 2023).

Third-party OTA booking sites, such as Expedia or Booking.com, can charge hotels a 15%-30% commission fee for each booking. This high acquisition cost is not sustainable as the OTAs are reaping the benefits of each booking and, in return, lowering the profitability of hotels. OTAs encrypt guest booking information; therefore, hotels do not even have so much as an e-mail address that they can use as a future marketing tool (Revenue-Hub, 2023). OTAs also benefit from increased brand awareness and brand loyalty, which the hotels could benefit from if they increased their direct bookings. Due to the perishable nature of hotel rooms, which cannot be saved for the future if unsold, hotels will always need an effective multi-channel distribution system. One way to ensure that hotels remain profitable in this digital economy is to increase their direct bookings (Barmpa, 2017). To understand what channels hotel bookings are made through and the distribution system, one must first understand the different categories that define the booking process. A thorough literature review is presented that will help shed light on this process.

9.2 Literature Review

9.2.1 Booking Channels

Organising bookings through booking channels is essential to all tourism businesses (Anwar et al., 2014) and is the foundation of the tourism business. Without booking channels, tourists would find it difficult to arrange travel services, flights, accommodation, and other travel-based services. Travel agents are one of the first booking channels that evolved in the tourism industry. Seočanac et al. (2019) and Morales (2017: 1) describe booking channels in the following way:

> There are two ways tourism products reach a customer, namely direct or indirect distribution channels. Direct distribution, such as through a website or phone, is when a product or service leaves the producer and goes directly to the customer. In contrast, indirect distribution is when intermediaries or middlemen are involved within the channel. For example, OTA and GDS are indirect distribution channels. Both direct and indirect distribution channels can be offline or online.

For this study, we refer to indirect channels as OTA bookings. Direct bookings are done via the hotel and/or its website. This study uses Banoobhai's (2017) conceptual framework to visually represent the relationship between different types of e-commerce (EC) in hospitality. The framework showcases the booking channel processes used within the hospitality industry; Airlines, Accommodation, Business-to-Business (B2B), and Business-to-Consumer (B2C) sales channels, as well as OTAs and merchants, all operate through the "World Wide Web" (WWW) commonly known as the Internet. Banoobhai's (2017) framework also shows how the WWW has been successful in the development of closed and open networks. The Internet is a virtual information system that has undeniably affected travel bookings (Fairer-Wessels, 2004).

Prior to the Internet's inception, most of these bookings were completed through intranets, connecting merchants with distributors and not being directly available to the public, namely Global Distribution Systems (GDS) like Amadeus, Sabre, and Travelport (which combines Apollo, Galileo, and Worldspan), which only distributors like Travel Agents and Hotels/Airlines had access to (Altexsoft, 2023).

The study also draws from Heuvel's (2022) conceptual framework which explains the booking process and the Property Management Systems (PMS) and technology that support said process, whether Direct or Third party. Direct bookings come via the hotel's website, telephonically, via email, or as walk-in guests. Hotels also receive corporate and government bookings, meetings, incentives, conferences, and events (MICE) markets. Third-party bookings come via OTAs and tour operators who sell rooms for the hotel at a cost. By removing the "middleman", hotels' cost of acquisition can decrease and thus lead to better sustainability and maximise gross profit (revenue minus distribution costs) rather than the income from a single room sale (Heuvel, 2022).

Greensill's (2016) South African study defines this process as "Disintermediation". It conducts the same argument as Heuvel, referring to the combined approach of looking for new distribution channels as well as "Reintermediation" for hotels to be profitable and not relying solely on third parties for distribution. Disintermediation means removing the middleman completely, whereas Reintermediation refers to forming new relationships with intermediaries offering value-added services to complement the ever-changing environment.

9.2.2 Advantages and Disadvantages of OTAs

The previous section discussed the various booking channels and booking process; further to these, there are advantages and disadvantages of these booking channels. OTAs such as Booking.com, Expedia, and Airbnb (Cloudbeds, 2023) have access to a broad customer base, and as such, hotels that use this distribution channel can reduce their marketing spend. Hotel Managers perceive this as simplifying the customer acquisition process and increasing brand awareness on these distribution channels (Bigne et al., 2021; Punnett, 2023; NI Business Info, 2023; Roma et al., 2014).

However, the disadvantages like high commission fees, loss of customer ownership, and the need for a multi-channel strategy put these advantages into question. OTAs are highly impersonal, and hotels can claim back these bookings by customers booking directly to regain control of their Customer Relationship Management (CRM) and invaluable Customer Data; further to this, Gazzoli et al. (2008) reported inconsistency regarding the availability of rooms which lead to unreliability. From a consumer perspective, pricing is the number one reason to book via OTAs, as they are perceived to have the lowest prices (Melis & Piga, 2017). This last sentiment is confirmed by Muresherwa et al. (2022) study in the Cape Metropole, which found that OTAs are the preferred method of booking rather than direct reservations mainly due to convenience and affordability. Table 9.1 presents a summary of the main advantages and disadvantages of OTAs.

9.2.3 Advantages and Disadvantages of Direct Bookings

In Table 9.2, Devisme (2023) remarks that hotels have the most to gain from direct bookings. This channel can provide the highest margin for revenue, enable the hotel to upsell their products, for example: "Stay three nights get one free", and retain all guest data, which can be used for future marketing purposes or following-up to convert "lookers" to "bookers" as well as control the guest experience from the research and booking process through to guest arrival, stay and departure.

Some of the main reasons hotel managers ignore the direct booking strategy could be increased costs, from marketing to PMS software and subscriptions, to building, developing, and maintaining the appropriate website and finding the right

Table 9.1 Advantages and disadvantages of OTA booking channels in the hotel industry

Advantages	Disadvantages
Access to a broader audience	High commission fees
Reduced marketing spends	A multi-channel strategy is still required.
Simplified process	Impersonal—Lack of CRM
Increased brand awareness	Loss of customer ownership—Data

Source: Bigne et al. (2021), Punnett (2023), NI Business Info (2023), Roma et al. (2014)

Table 9.2 Advantages and disadvantages of direct bookings

Advantages	Disadvantages
Highest margin	Cost—Marketing, technology website
Upselling techniques	Talent
Ownership of consumer data	OTA domination
Control guest experience	Inertia or lack of understanding

Source: Devisme (2023), Patel (2022)

talent to manage these complex systems. Some could be daunted by the dominant nature of OTAs in the tourism and hospitality industry and not see another solution, or they are stuck in inertia from a lack of understanding of the booking process (Patel, 2022).

Mndzebele's (2013) South African study mentions six advantages of direct booking channels using B2C via EC. They can be summarised as follows: [1] EC needs to be adopted, managed, and operated at maximum efficacy if businesses want to remain competitive, [2] EC is an opportunity to create new relationships with consumers to build trust and loyalty, [3] enables direct contact with the business, [4] and eliminates costs of intermediaries (third parties), [5] consumers can reduce their buying risk by having more information relating to the product/service they are buying and lastly [6] increase in B2C transactions will increase consumer loyalty and trust, meaning hotels need to invest in the most recent technologies for their websites.

9.2.4 Strategies to Win Hotel Direct Bookings

Guests' decision-making on hotel bookings is a complex behaviour and a challenge to narrow down all the factors contributing to customer decision-making; therefore, several strategies will be mentioned (Amaro & Duarte, 2013). The hotel website is an essential primary tool that markets services at a reasonable cost (Anwar et al., 2014). In Khumalo-Ncube & Motala's (2021) South African study, an emphasis is placed on hotel booking website quality. High-quality websites tend to attract more customers and are a critical success factor in purchase intentions. Another South African study by de Jager and Ezeuduji (2015) mentions trust, convenience, and price as mitigating factors customers will look for the most perceived value. Heuvel (2022) advises that hotels' booking strategies rely on how they position themselves in the market to differentiate between themselves and OTA's, such as leisure (pool and spa facilities) or business travel (conferencing and Wi-Fi) and thus need to implement different strategies to capture their target market who prefer specific hotel attributes over others.

Masiero and Law (2016) also mention price, promotional offers, and the booking website as significant factors to consider in the consumer booking process, and hotel websites should focus on these elements if they wish to garner the attention of prospective guests. Gazzoli et al. (2008) mention that price disparities are one of the main reasons consumers shop around. This is an excellent argument for rate parity and a proper marketing strategy, which Heuvel (2022) suggests. Lee et al. (2013) further confirm that conflicts can occur when OTAs do not have rate parity with hotels and offer dynamic pricing. Garrido-Moreno et al.'s (2021) recent study on COVID-19 has demonstrated the usefulness of the following strategic marketing measures in business recovery: develop specific campaigns and offers to address the local market; develop and promote new products and services (to generate alternative revenue); extend loyalty program benefits; prepare promotional plans to increase future stays; update and improve channels to contact, attract, and sell to customers; digitalise marketing strategies and advertise through different media channels.

Mouakket and Al-Hawari (2012) agree that improving hotels' marketing strategies will increase customer retention. In addition to website quality, rate parity, and marketing strategies, Jun-Joe (2014) states that value adds are increasingly crucial to hotels in managing their inventory. Customer Relationship Management can build trust by introducing loyalty programs (Dickinger & Mazanec, 2008; Park et al., 2018). Heuvel (2022) states that hotels can maintain the upper hand if [1] they offer the best prices on their Hotel Website which will encourage direct bookings, [2] together with the considered use of a digital marketing strategy like social media presence, [3] Search Engine Optimization (SEO) and [4] pay-per-click marketing as well as the [5] website using the correct booking technology; can all work together to create a more suitable environment to promote direct bookings. While OTAs remain a vital part of the industry and show no signs of retreat, hotels can use their OTA presence to drive traffic to their leading websites. A little investment into their direct booking strategy and marketing strategy would go a long way to encourage direct bookings (Hollander, 2023).

9.3 Research Methodology

9.3.1 Study Setting

The study was conducted in the Cape Metropole in Cape Town. The Cape Metropole is situated in the southern peninsula of Western Cape Province. It stretches 294 km from Gordon's Bay to Atlantis along the coastline, covering 44 cities (South Africa Tourism, 2023; Municipalities of South Africa, 2023). Most hotels are in Cape Town and the Southern Suburbs, as shown in Fig. 9.1. The figure on the right shows the areas within the Cape Metropole, and the one on the left shows the location of the hotels that participated in the study.

Fig. 9.1 Cape Town Metropole. (Municipalities of South Africa, 2023)

9.3.2 Methodological Approach

The cross-sectional design was conducted over a three-month (April–June) period in 2021. The target population was four- and five-star in Cape Town, with respondents being the management who formed part of the decision-makers and worked directly with the technology systems. The Tourism Grading Council of South Africa (TGCSA, 2016) was used to determine the sampling frame for this study. When the survey was conducted, TGCSA showed 63 four- and five-star hotels in the Cape Metropole. The study was conducted during COVID-19, and nine of these 63 hotels had closed (Abbas et al., 2021), so the population was reduced to 54 hotels. Non-probability, convenience, and purposive sampling (Rahi, 2017) were used to select hotels and managers. Thirty-eight (38) hotels participated in the study, and a manager from each hotel completed the questionnaire.

A quantitative approach was applied using an online questionnaire in Google Forms. The questionnaire items developed were guided by the literature on direct bookings, OTA bookings, and guest booking choices. The online questionnaire was self-administered, and then the data was downloaded on the Microsoft Excel Spreadsheet, coded, screened, and imported to SPSS Version 26 for analysis. The analysed data was presented in table format using descriptive statistics such as percentages, frequencies, average, and standard deviation.

Ethical protocols were followed to ensure fair treatment of the participants. To ensure this, the study received ethics approval from the institution, permission from the hotel management, consent from the respondents, an introduction explaining the purpose of the study, and the ethical aspects to be followed during data collection, analyses, and presentation of the results. Volunteerism, withdrawal, anonymity, confidentiality, and omission were the aspects considered, to name a few.

9.4 Results and Discussion

In response to the aim of the study, hotel managers were asked specific questions to understand the eagerness and challenges they face when trying to attract guests to book directly with them instead of using third parties (OTAs). Therefore, this section discusses these aspects. In addition, the section first presents and discusses the background information of the hotels that participated in the study. Fifty-four hotels were approached to participate in the study; however, 38 hotels consented to participate, and a manager from each hotel completed the questionnaire.

9.4.1 Hotel Background Information

Hotel grading influences the level of service quality expectations and customer perception, affecting loyalty and, therefore, booking. Customers would prefer to book higher-graded hotels in anticipation of receiving good quality service (Tefera & Govender, 2015). Chivandi et al. (2020) findings support this statement as most of the respondents indicated that their last hotel booking was either at four- or five-star hotels. Therefore, the anticipation was that four- and five-star hotel management would be fit to respond to the purpose of the study; hence, a fair representation of four- (52.6%) and five-star hotels (47.4%) participated in the survey (TGCSA, 2016).

Branding is an essential aspect of hotels regarding identity, marketing, and economic contribution, as these influence guests' offerings and choices (Ribaudo et al., 2020). Most managers (68.4%) indicated that the hotels participating in the study were identified as hotel chains, and 31.6% were independent hotels (Table 9.3).

Managers held various positions, with most being the General Managers (44.7%), followed by Revenue Managers (28.9%) and the owner (10.5%). These findings concur with Banoobhai's (2017) findings, where hotel managers who participated in the study held various management-level positions (Fig. 9.2).

Close to half of the travel purchases are done through Online Travel Agents (OTAs) (Mouakket & Al-Hawari, 2012), and these are well-known as the key distribution channels for many hospitality businesses (Melis & Piga, 2017). This is observed in this study, whereby 52.6% of the managers say that OTA is the predominant channel guests use to book hotels. In line with the literature mentioned earlier, these results imply that 52.6% of the hotels pay 15–30% commission to OTAs (Guo et al., 2020). Anwar et al. (2014) attest to this as they have observed that some of the OTAs portals' commissions are very high. Although that is the case, 29% (email at 16%, website at 11%, and walk-in at 3%) indicated guests predominantly use direct booking. This is not unique to this study, as Schegg et al. (2013) reported that guests use mainly direct booking channels when reserving accommodation. This is further in line with the findings by Anwar et al. (2014), who said that 37% of accommodation used direct booking platforms (Refer to Table 9.4).

Table 9.3 Hotel grading and categories

Hotels information		
	Frequency	%
Hotel grading		
Four-star	20	52.6%
Five-star	18	47.4%
Total	38	100%
Hotel categories		
Chain hotels	26	68.4%
Independent hotels	12	31.6%
Total	38	100%

Fig. 9.2 Hotel management who completed the questionnaire

Table 9.4 Hotel booking channels

Booking channel	Frequency	%
OTA	20	52.6
Email	6	15.9
Own Website	4	10.5
Tour Operator	4	10.5
Telephone	3	7.9
Walk-in	1	2.6
Total	38	100

9.4.2 Strategies to Win Direct Booking

The results show hotel managers (98%) are striving to grow direct bookings. Managers were asked if direct booking is one of their strategic focuses, and the majority (98%) indicated that they want to win and grow direct bookings. When reviewing the level of direct hotel bookings, the results show that only 10.5% of hotel bookings were made through the hotel's website (Fig. 9.3).

In line with what the managers said above regarding their striving to win guests through direct booking, managers were asked certain questions to indicate their strategies to attract and keep guests. The managers said the hotel uses loyalty programmes (57.9%) and incentives (68.4%). Loyalty programmes are one of the strategies hotels implement to encourage repeat business and drive direct bookings (Koo et al., 2020). La Rose and Rowson (2017) define its benefits as strengthening the relationship with the guest by repeat bookings directly with the hotel. Encouraging guests to book now via the hotel website can be challenging for hotels, as guests review multiple online platforms, searching for the best deal (Emir et al., 2016). If hotels display the same rates across online booking platforms, offering guests an

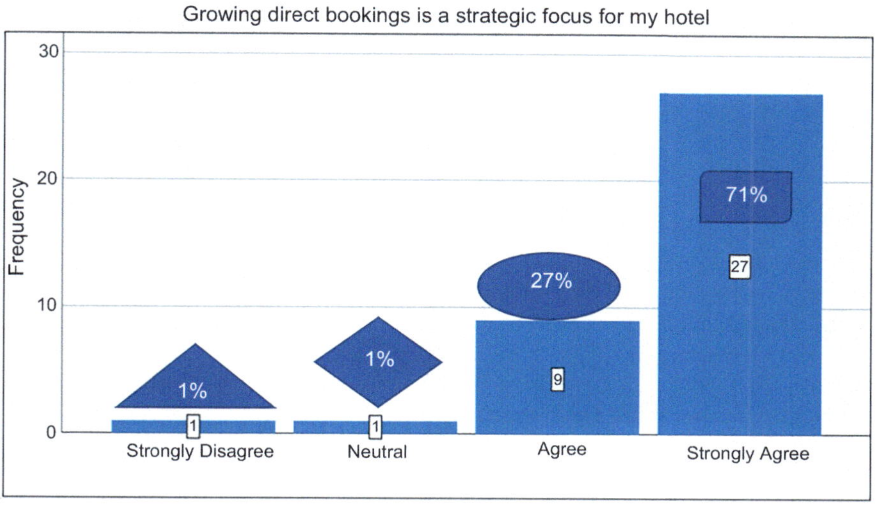

Fig. 9.3 Growing direct booking is a strategic focus for my hotel

incentive to book via the hotel website could be a significant factor in the guests' decision-making process (Fig. 9.4).

Although 98% of the managers indicated that direct booking is their strategic focus and most of the hotels offer loyalty programmes (57.9%) and incentives (68.4%), it is concerning that more than 30% of the hotels are not using loyalty programmes (42.1%) and incentives (31.6%) as a strategy to attract and retain return guests through direct booking and increase brand loyalty (Ye et al., 2019; Jun-Joe, 2014) which yield to long term profit (Mouakket & Al-Hawari, 2012). This is confirmed by Independent Hotel Show (2019), whereby 70% of hotels offered incentives to guests to encourage direct bookings. These hotels are, therefore, losing an opportunity to promote business repeat via direct booking channels. Managers from these hotels were asked if there were reasons for them not to have loyalty programmes and incentives in place, and these are presented in Table 9.5. For this reason, there are clear challenges that hotels are facing, causing them to have neither loyalty programmes nor incentives, such as budget/financial and time constraints, the aftereffect of COVID-19, technological constraints, and OTAs obligations and offerings. Literature attests that loyalty programs have cost implications as managing the program incurs more expenses (Jun-Joe, 2014).

9.4.3 Challenges to Win Direct Bookings

Even though the hotels are striving to win guests through direct booking, several challenges may hinder this strategy. The first challenge is that they do not know the acquisition cost for direct booking (76.3%). These results allude to Merkert and

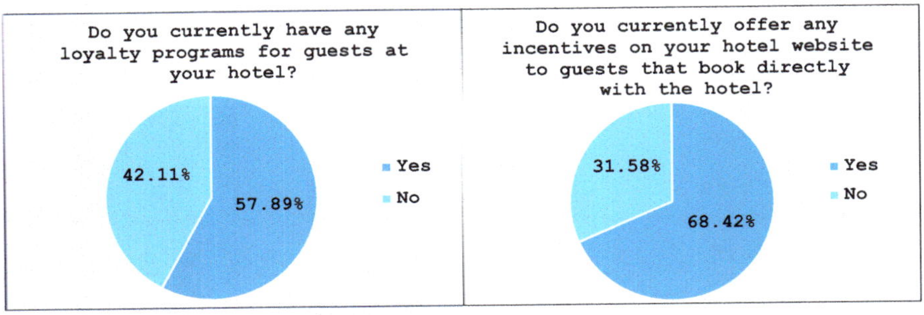

Fig. 9.4 Hotels strategies to win direct booking guests

Table 9.5 Reasons for not having a loyalty programme and incentives

Reasons for no strategies to attract and retain guests			
	No loyalty programme		No incentive programme
1	Working on a local membership and then an overseas membership	1	We offer packages
2	A mix of budget and time constraints	2	OTA contractual obligations
3	A combination of limited budget and time constraints	3	Not able to offer special discounts on the website yet
4	I have not gotten around to it	4	Financial constraints due to COVID-19
5	I do not feel loyalty programs drive more business. Instead, we focus on delivering individualised gifting and value-added services to build value	5	Not sure
6	The group is too small to offer a reward system	6	Guests are encouraged to join the rewards program
7	We offer return guests value add and discounts	7	Radisson Rewards members will receive preferential rates as part of the loyalty program
8	We offer return rates/discounts to regular guests	8	Respect for global revenue generators
9	Small, independent hotel	9	BAR rates are already offered
10	We are small; we look after our regulars on a personal level, each one different	10	Rate integrity is essential to us, and although we pay commission to OTAs, the marketing exposure and additional visibility we receive in return compensate for this
11	Before COVID-19, the hotel ran at over 90% occupancy for the year	11	The channel manager can only offer one discount. Should a client see a better discount on OTA, this cannot be afforded through the website, and the client will have to contact the hotel directly to receive further discounts
12	Future plans		
13	I am still looking for the correct program to use		
14	We do not currently have a rewards/points system as a hotel group		
15	We do not have a global rewards program yet		

Hakim (2022), who stated that transactional costs impact the business's effectiveness in the tourism sector, yet the extent of these costs is scant. These costs assist hotel management in understanding the financial implications associated with direct booking versus OTAs. Meanwhile, OTAs consume more than 40% of their rate, which may lead to a loss of revenue for hotels (Revenue Hub, 2023). Jun-Joe (2014) stated that intermediary fees cost more than USD 2.5 billion and are anticipated to increase. Revenue Hub (2023) says that acquiring direct booking is less costly than OTAs and more profitable, with an average 12.5% profit. This was further attested by Anwar et al. (2014), stating that a business website is a cost-effective tool to display the service offered and engage with guests.

The second challenge is that although hotel management considers the hotel website to be the most crucial channel for direct booking (Teng et al., 2020), hotels do not have control over their website, and this may affect their decision-making capabilities. Most of the hotels' websites are managed externally (74%) by either head office (42%), outsourced suppliers (29%), or globally (3%) (Refer to Table 9.6).

The third main challenge is that most guests book via OTAs (refer to Table 9.4) instead of directly with the hotels. Lien et al. (2015) mentioned that guests do so because of time-saving, cost-effectiveness, and convenience. The introduction of the Internet in South Africa brought a shift in consumer buying behaviour. Hotels observed a change in the booking method whereby customers are using multiple booking channels, mainly OTAs, and this brought a challenge to hotels because customers preferred OTAs as they offered various benefits such as affordability, convenience, time, price, and other benefits such as transfer from the airport, flight booking (Muresherwa et al., 2022; de Jager & Ezeuduji, 2015). The abovementioned concur with this study as managers indicated that guests use OTAs because of convenience, variety of accommodation establishments, accessibility, flexibility, and the security of OTA's booking technologies, highlighting the most prominent

Table 9.6 Hotel direct booking cost, channels, and website management

Direct booking cost, channels, and website management					
Knowledge of the cost of acquisition for direct booking					
	N	%			
No	29	76.32			
Yes	9	23.68			
Channels for most direct booking					
N (%)					
Strongly agree	Agree	Neutral	Disagree	Strongly disagree	Total
20 (53)	8 (21)	7 (18)	1 (3)	2 (5)	38 (100)
Hotel website management					
Internally by dedicated personnel	By head office	By an outsourced supplier	**Globally**		Total
10 (26)	16 (42)	11 (29)	1 (3)		38 (100)

Fig. 9.5 Reasons managers feel guests prefer to book via OTAs

ones. Masiero et al. (2020) add that OTAs offer customers benefits such as free cancellation and payment flexibility as guests can pay upon arrival. A study by Jun-Joe (2014) observed that OTAs offer immediate loyalty programs; in contrast, hotels give delayed rewards, giving OTAs a competitive advantage as customers prefer instant. These OTAs also offer visibility, essential in attracting and retaining guests (Anwar et al., 2014). Refer to Fig. 9.5.

9.5 Conclusions

This study identified background information that attests that hotels are eager to attract guests for direct booking. On the other hand, there are barriers to be overcome which are hindering hotels in the Cape Town Metropole from winning direct bookings. The study contributes to the understudied and current phenomenon of understanding hotels' challenges to win guests for direct bookings. Therefore, this study presents limited and valuable information contributing to a body of knowledge about hotels and the accommodation industry at large. This study's focus was achieved using a quantitative approach, closed and few open-ended questions, whereby managers shared their views on winning direct bookings and established aspects that may hinder the realisation of winning guests.

Managers indicated that hotels are willing to win back direct booking, which is part of their strategic focus to the extent that loyalty programmes are in place to attract and retain guests. However, several challenges may hinder the actualisation of direct bookings. The first challenge is that hoteliers do not know the acquisition cost for direct booking, which is important for decision-making regarding the best cost-effective approach for hotel bookings. The second challenge is that they do not have control over their website, which is a valuable tool for direct bookings, marketing, and distribution. Hotels must gain control over their websites to enhance their usability and make them more efficient and enjoyable for potential customers (Shasha, 2013). Hotels must consider investing in human resources dedicated to managing the direct booking systems internally and encouraging human interaction through the online direct bookings system and other media such as mobile and social network platforms to enhance the personalised experience.

On the other hand, the third challenge is that guests prefer to book accommodation through OTAs because of its various benefits. Hotels must introduce competitive benefits to match OTA's offering (Jun-Joe, 2014). Furthermore, hotels should invest in moving with the trends to keep up with OTAs and remain competitive. To do so, they need to consider accessing mobile and social media network users as these platforms have grown. These influence consumers' habits and ensure presence on the websites which book corporate customers (Mouakket & Al-Hawari, 2012). High-quality websites tend to attract more customers and are a critical success factor in purchase intentions. Website quality influences purchasing intentions as it strongly relates to customer satisfaction (Khumalo-Ncube & Motala, 2021).

There is a growing trend worldwide about winning back hotel accommodation direct bookings. Still, there is a minimal scholarly contribution in this regard; furthermore, the aspects that hinder this from being actualised. The literature identified examined the benefits from the customers' perspective, mainly looking at customer behaviour (Chubchuwong, 2022; Muresherwa et al., 2022; Mugica & Berne, 2020; Ye et al., 2019). The lack of literature regarding direct hotel bookings prompts the focus of this study to understand direct booking and its challenges from the managers' perspective. The paucity of literature became a limitation for this study as it took time to find sources to corroborate the findings of this study. Therefore, this strengthens the contribution of this study in academia by adding to scarce literature and presenting findings that will assist hotel management in understanding direct booking challenges and then finding solutions to the challenges.

References

Abbas, J., Mubeen, R., Terhemba, P., & Raza, S. (2021). Current research in behavioural sciences exploring the impact of COVID-19 on tourism: Transformational potential and implications for a sustainable recovery of the travel and leisure industry. *Recent Research in Behavioural Sciences, 2*(February), 100033.

Altexsoft. (2023). *How to choose a GDS: Amadeus vs Sabre vs Travelport*. https://www.altexsoft.com/blog/travelport-vs-amadeus-vs-sabre-gds/ Accessed 3 Sept 2023.

Amaro, S., & Duarte, P. (2013). Online travel purchasing: A literature review. *Journal of Travel & Tourism Marketing, 30*(8), 755–785.

Anwar, M. A., Carmody, P., Surborg, B., & Corcoran, A. (2014). The diffusion and impacts of information and communication technology on tourism in the Western cape, South Africa. *Urban Forum, 25*, 531–545. https://doi.org/10.1007/s12132-013-9210-4

Banoobhai, I. (2017). *The role of e-commerce in four-and-five-star Hotels in the Cape Town Metropole*. Master's Thesis, Cape Peninsula University of Technology.

Barmpa, A. (2017). *How can a hotel increase its direct bookings? The case study of a Hotel in Bulgaria*. International Hellenic University. Dissertation.

Bigne, E., Nicolau, J. L., & William, E. (2021). Advance booking across channels: The effects of dynamic pricing. *Tourism Management, 86*, 104341.

Callan, R. J. (1995). Hotel classification and grading schemes, a paradigm of utilisation and user characteristics. *International Journal of Hospitality Management, 14*, 271–284. Pergamon.

Chivandi, A., Samuel, O., & Muchie, M. (2020). The effect of Online Travel Agencies (OTA) on brand relationships in low and high-end hotels in South Africa. *Geojournal of Tourism and Geosites, 31*(3), 951–957. https://doi.org/10.30892/gtg.31303-52

Chubchuwong, M. (2022). Factors affecting online direct hotel booking: A study in Bangkok. *APHEIT International Journal, 11*(1), 32–47.

Cloudbeds. (2023). *Top 10 Online Travel Agencies in 2023*. https://www.cloudbeds.com/online-travel-agencies/best. Accessed 21 Aug 2023.

de Jager, K., & Ezeuduji, I. O. (2015). Socio-demographic variables' relationship in choosing between travel agencies and the Internet for leisure travel arrangements: The case of South Africa. *African Journal of Hospitality, Tourism and Leisure, 4*(2), 2223-814X.

Devisme, B. (2023). *How to increase your hotel direct bookings in 2023: two strategies most hotels are missing*. https://www.quicktext.im/blog/strategies-to-increase-hotel-direct-bookings-in-2023/ Accessed 3 Sept 2023.

Dickinger, A., & Mazanec, J. (2008). *Consumers' preferred criteria for hotel online booking* (pp. 244–254). https://doi.org/10.1007/978-3-211-77280-5_22.

Emir, A., Halim, H., Hedre, A., Abdullah, D., & Azmi, A. (2016). Factors influencing online hotel booking intention: A conceptual framework from stimulus-organism-response perspective. *International Academic Research Journal of Business and Technology, 2*(2), 129–134.

Fairer-Wessels, F. A. (2004). Web-based information systems: The development of Leisuredigest.com – A South African tourism website. *Mousaion, 22*(1), 52–69.

Garrido-Moreno, A., García-Morales, V. J., & Martín-Rojas, R. (2021). Going beyond the curve: Strategic measures to recover hotel activity in times of COVID-19. *International Journal of Hospitality Management, 96*, 102928.

Gazzoli, G., Kim, W. G., & Palakurthi, R. (2008). Online distribution strategies and competition: Are the global hotel companies getting it right? *International Journal of Contemporary Hospitality Management, 20*(4), 375–387.

Greensill, G. (2016). *Understanding the international consumers' behaviour towards disintermediation in the luxury hotel market*. Unpublished Master's Thesis, GIBBS University of Pretoria.

Guo, Q., Anderson, C. K., Dong, J., Zhao, P., & Ji, Q. (2020). Coordination contracts for hotels and online travel agents. *Sustainability (Switzerland), 12*(8), 1–17.

Heuvel, D. (2022). *Hotels' strategic objectives regarding direct bookings at four-and-five-star hotels in the Cape Town Metropole*. Master's Thesis. Cape Peninsula University of Technology.

Hollander, J. (2023). *Hotel Direct Bookings: Breaking down how hotels can win the OTA battle in 2023*. https://hoteltechreport.com/news/hotel-direct-bookings. Accessed 28 Aug 2023.

Independent Hotel Show. (2019). *The relationship between hotels and booking sites!* [Online]. https://www.independenthotelshow.nl/news/66-the-relationship-between-hotels-and-booking-sites. Accessed 15 Sept 2023.

Jun-Joe, S. (2014). *The effect of loyalty program attributes on customer's booking choice*. Master's thesis, Purdue University.

Khumalo-Ncube, S., & Motala, T. (2021). Hotel booking website quality, travel agent satisfaction and purchase intention. *African Journal of Hospitality, Tourism and Leisure, 10*(6), 1932–1943.

Koo, B., Yu, J., & Han, H. (2020). The role of loyalty programs in boosting hotel guest loyalty: Impact of switching barriers. *International Journal of Hospitality Management, 84*, 102328.

Muresherwa, G., Kyomba, M. N., & Spencer, J. P. (2022). Indirect versus direct bookings: Hotel customer motivations for online reservations at travel agencies in the Cape Metropolis. *Acta Universitatis Danubius, 18*(5), 2022.

La Rose, F., & Rowson, B. (2017). The benefit of guest loyalty programmes. *Research in Hospitality Management, 7*(1), 11–15.

Lee, H. A., Denizci, G. B., & Law, R. (2013). An examination of the relationship between online travel agents and hotels: A case study of Choice Hotels International and Expedia.com. *Cornell Hospitality Quarterly, 54*(1), 95–107.

Lien, C., Wen, M., Huang, L., & Wu, K. (2015). Online hotel booking: The effects of brand image, price, trust, and value on purchase intentions. *Asia Pacific Management Review, 20*(4), 210–218. https://doi.org/10.1016/j.apmrv.2015.03.005

Liu, X., Li, C., Nicolau, J. L., & Han, M. (2023). The value rating diversity within a multidimensional rating system: Evidence from a hotel booking platform. *International Journal of Hospitality Management, 110*, 103434.

Masiero, L., & Law, R. (2016). Comparing reservation channels for hotel rooms: A behavioral perspective. *Journal of Travel and Tourism Marketing, 33*(1), 1–13.

Masiero, L., Viglia, G., & Nieto-Garcia, M. (2020). Strategic consumer behavior in online hotel booking. *Annals of Tourism Research, 83*, 1–31.

Melis, G., & Piga, C. A. (2017). Are all online hotel prices created dynamic? An empirical assessment. *International Journal of Hospitality Management, 67*, 163–173.

Merkert, R., & Hakim, M. M. (2022). Travel agency transaction costs in airline value chains – A risk in distribution channels in South Asia? *Annals of Tourism Research, 95*, 103414.

Mndzebele, N. (2013). The adoption of EC in B2C order-taking in the Hotel Industry. *International Journal of Advanced Computing Technology., 2*(2), 82–85.

Morales, T. (2017). *Increasing direct bookings in hotels*. Haaga-Helia University of Applied Sciences. Dissertation.

Mouakket, S., & Al-Hawari, M. A. (2012). Examining the antecedents of e-loyalty intention in an online reservation environment. *Journal of High Technology Management Research, 23*(1), 46–57.

Mugica, J., & Berne, C. (2020). Direct and indirect tourism online channels. Do they have a different potential for customer loyalty? *Sustainability, 12*(11), 1–15.

Municipalities of South Africa. (2023). *City of Cape Town Metropolitan Municipality (CPT)*. https://municipalities.co.za/map/6/city-of-cape-town-metropolitan-municipality. Accessed 19 Aug 2023.

Naidoo, J. (2022). *Hello Darlings Travel Scandal: 10 things you should know*. https://www.iol.co.za/news/south-africa/hello-darlings-travel-scandal-10-things-you-should-know-e11cecd3-8c84-4327-bb94-28232c84ee37 Accessed 3 Sept 2023.

NI Business Info. (2023). *Advantages and disadvantages of online travel agents*. https://www.nibusinessinfo.co.uk/content/advantages-and-disadvantages-online-travel-agents. Accessed 21 Aug 2023.

Park, S., Yin, Y., & Son, B. G. (2018). Understanding of online hotel booking process: A multiple method approach. *Journal of Vacation Marketing*. https://doi.org/10.1177/1356766718778879

Patel, C. (2022). *Is the direct booking strategy still worth pursuing?* https://www.linkedin.com/pulse/direct-booking-strategy-still-worth-pursuing-chetan-patel/?trk=pulse-article. Accessed 3 Sept 2023.

Punnett, E. (2023) *The advantages and disadvantages of online travel agencies*. https://www.ontarioseo.ca/blog/advantages-disadvantages-online-travel-agencies. Accessed 21 Aug 2023.

Rahi, S. (2017). Research design and methods: A systematic review of research paradigms, sampling issues and instrument development. *International Journal of Economics & Management Science, 6*(2), 1000403.

Rall, S. (2023). *Travellers' dreams dashed after agency cancels Maldives holiday, allegedly vanishes with monies*. https://www.iol.co.za/news/south-africa/kwazulu-natal/travellers-dreams-dashed-after-agency-cancels-maldives-holiday-allegedly-vanishes-with-monies-6564bc34-5e49-40b1-984e. Accessed 3 Sept 2023.

Revenue-Hub. (2023). *Rising guest acquisition costs and the value of direct booking*. https://revenue-hub.com/guest-acquisition-costs-value-direct-booking. Accessed 29 Aug 2023.

Ribaudo, G., Moccia, S., Orero-Blat, M., & Palacios-Marqués, D. (2020). Comparing chains versus independent hotels based on international sales: An exploratory study. *Economic Research-Ekonomska Istraživanja., 33*, 2286. https://doi.org/10.1080/1331677X.2019.1710719

Roma, P., Zambuto, F., & Perrone, G. (2014). Price dispersion, competition, and the role of online travel agents: Evidence from business routes in the Italian airline market. *Transportation Research Part E, 69*, 146–159.

Schegg, R., Stangl, B., Fux, M., & Inversini, A. (2013). Distribution channels and management in the Swiss Hotel Sector. In *Information and communication technologies in tourism 2013. Conference paper* (pp. 554–565).
Seočanac, M., et al. (2019). Online booking versus travel agents: Perspective of young tourists in Serbia. *Hotel and Tourism Management, 7*(2), 105–114.
Shariffuddin, N. S. M., Azinuddin, M., Yahya, N. E., & Hanafiah, M. H. (2023). Navigating the tourism digital landscape: The interrelationship of online travel sites' affordances, technology readiness, online purchase intentions, trust, and E-loyalty. *Heliyon, 9*, e19135.
Shasha, Z. T. (2013). User testing of Cape Town hotel reservation systems usability. In *Conference: The 15th World Wide Web conference*. Available https://www.researchgate.net/publication/321359007_User_testing_of_Cape_Town_hotel_reservation_systems_usability_0052. Accessed 15 Sept 2023.
South Africa Tourism. (2023). *Big City Life: Cape Town*. Available https://www.southafrica.net/gl/en/travel/category/places-to-go/big-city-life/cape-town. Accessed 15 Sept 2023.
Tefera, S., & Govender, K. (2015). Hotel grading, service quality, satisfaction and loyalty—Proposing a theoretical model and relationship. *African Journal of Hospitality, Tourism and Leisure, 4* (Special edition), 1–17.
Teng, Y. M., Wu, K. S., & Chou, C. Y. (2020). Price or convenience: What is more important for online and offline bookings? A study of a five-star resort hotel in Taiwan. *Sustainability (Switzerland), 12*(10), 1–16.
Tomlinson, Z., Venske, E., & Mokoena, P. (2023). Online response management: Hotelier perspectives from a south African tourist destination. *Studia Periegetica, 1*(41), 153.
Tourism Grading Council of South Africa (TGCSA). (2016). *Graded establishments*. https://www.tourismgrading.co.za/find-a-graded-establishment/search-for-graded-accommodation/ Accessed 15 Sept 2023.
Ye, F., Yan, H., & Wu, Y. (2019). Optimal online channel strategies for a hotel considering direct booking and cooperation with an online travel agent. *International Transactions in Operational Research, 26*(3), 968–998.

Part IV
Policy and Practical Recommendations

Chapter 10
Towards Inclusive Digitalisation in Small Hospitality and Tourism Enterprises

Emmanuel Ndhlovu, Kaitano Dube, and Ngoni Courage Shereni

10.1 Introduction and Background

Digitalisation drives the adoption of digital technologies, transforming businesses and business environments. It transforms markets and production processes and has significant consequences for economic and social organisation, innovation, and competitiveness (Kindzule-Millere & Zeverte-Rivza, 2022; Marx et al., 2021; Ndhlovu & Dube, 2023). In tourism, digitalisation presents opportunities for enterprises to expand their markets, increase growth, improve operational competencies, and refine their competitive edge. Digitalisation offers the tools and structures to create and/or add value to tourism products and experiences (Dredge et al., 2018). However, the success of digitalisation hinges on the aptitude of the tourism industry to share, learn, and collaborate (Kusuma et al., 2020). At a collective level, digitalisation can also aid in developing and modifying product offerings, improve connectivity in destinations, generate data which can be used to track performance, and help to improve destination management (Opute et al., 2020; Troshani et al., 2018; van Dyk & Van Belle, 2019). The potential of digitisation to ensure the profitability of businesses, however, depends on how it is used to improve products and services to satisfy customers. Digitalisation also needs to be adopted to enable all categories of stakeholders to access and utilise it. It has to allow customers to share information and develop supportive mentoring opportunities for employees to help them learn, reflect, and grow.

Studies on digitalisation mostly dwell on the benefits and challenges (Filimonau & Naumova, 2020; Gokkaya, 2020; Hassan, 2023; Ndhlovu & Dube, 2023), with

E. Ndhlovu (✉) · K. Dube
Vaal University of Technology, Vanderbijlpark, South Africa
e-mail: kaitanod@vut.ac.za

N. C. Shereni
University of Johannesburg, Johannesburg, South Africa

E. Ndhlovu et al. (eds.), *Tourism and Hospitality for Sustainable Development*,
https://doi.org/10.1007/978-3-031-63069-9_10

very few recommendations made on how the process can be made as inclusive as possible to benefits different types of stakeholders. For instance, a major concern relates to how small businesses, particularly those in developing countries and rural areas, can be empowered to tap into the benefits of digitalisation. Studies have shown that enterprises located within developing countries and those located in rural areas experience several challenges with digitalisation when compared to enterprises in developed countries and those in urban areas, generally (Alawamleh et al., 2022; Amornkitvikai et al., 2022). Scholars found financial, education and training, organisational, and infrastructural factors to be some of the major challenges faced by enterprises in developing countries (Ndhlovu & Dube, 2023). Thus, studies which explore digitalisation challenges and obstacles, types of policy responses, and interventions in digitalisation are scant. A study covering these research aspects is essential to enable digitalisation in the different parts of the tourism industry and stakeholder groups. This chapter closes this gap by identifying digitalisation challenges faced by small tourism businesses and exploring strategies which may be used to promote inclusive technology adoption and integration by small tourism enterprises globally.

The chapter is organised as follows: The next section outlines the research methodology for the study. This is followed by a review of the literature on digitalisation in the tourism industry, digital tools in tourism and hospitality, and the factors affecting digitalisation in the industry. The chapter then discusses the strategies that may be used to improve digitalisation by enterprises in the tourism industry. Focus is placed on small enterprises. Lastly, conclusions are drawn from the discussion.

10.2 Materials and Methods

The chapter used a conceptual research approach to draw from existing knowledge and practical examples to highlight how technology adoption can be promoted inclusively. Conceptual research does not involve conducting any practical experiments. It is related to abstract concepts or ideas (Skoczylas, 2019). It typically gives the researcher an idea of the expected outcome. This guides the research, prevents mistakes, and helps the researcher focus on the project goals (Vaismoradi & Snelgrove, 2019). By utilising the conceptual approach, the authors were able to identify the challenges of digitalisation by small tourism businesses and to identify the strategies that can be used to support the digitalisation campaign by these businesses. The approach also enabled the authors to benefit from previous studies since it is largely based on studies conducted previously. The approach saved both time and resources since there were no fieldwork trips made to collect empirical data. The flexibility of the approach also enabled the researchers to select texts that would be considered for analysis without undergoing the complex and strict specifications often associated with qualitative and quantitative empirical research approaches. The flexibility of the approach also enabled the incorporation of the sectorial expertise of the authors to propose how the digitalisation process in small tourism enterprises can be fast-tracked.

The conceptual research approach, however, generates speculations that are based on existing writing instead of experimentation and perception and draws conclusions that are less fact-based and may not essentially be considered dependable. The authors believe that since digitalisation is still an unfolding process; conceptual studies are needed to provide the basis upon which ongoing research can be built. Future studies, should, therefore, problematise the findings of this study using empirical evidence.

10.3 Presentation and Discussion of Findings

10.3.1 Digitalisation in Tourism

The advent of digital technologies opened up many opportunities but also challenges for businesses. The digitalisation process helps businesses unlock the full potential of markets and operate more effectively to meet customer needs (Bigliardi et al., 2021). Digitalisation emerged with the advent of the Fourth Industrial Revolution (4IR). Fourth Industrial Revolution is the term used to describe many developments taking place in industries driven by digitalisation, including automation, interconnectivity between cyber and physical systems, and improved management competencies provided by big data analytics and cloud computing (Mhlanga et al., 2021; Mhlanga & Ndhlovu, 2023a). Digitalisation first became popular in the manufacturing industry but then quickly spread to other parts of the economy. The 4IR was accompanied by Business 4.0, transforming business practices through digitalisation and open innovation. According to Dredge et al. (2018), tourism, which carries the characteristics of both the industry and the service industry, is going through an enormous transformation due to the developments in information technologies and processes of digitalisation.

With a particular focus on the tourism industry, scholars identify three main stages of technology development. The first stage is related to sales and marketing. This stage roughly ran from the year 1990 to 2000. It was characterised by the introduction of the Internet, which assisted destination businesses in utilising technology as a marketing tool (Schallmo et al., 2017). The advent of digital point-of-sale and other supporting software helped tourism businesses to improve internal operations. During this stage, websites replaced paper-based marketing material; destination management businesses became 'information brokers'; online systems expedited business reservations and payments, and distribution systems eased coordination across the industry (Zeqiri et al., 2020).

The second stage ran from around 2000 to around 2010. Scholars describe this stage as the 'digital business ecosystems' stage (Dredge et al., 2018). During this stage, the Internet consolidated its position as the primary source of information for tourists (Shereni & Chambwe, 2020). Developments in Internet technologies facilitated the emergence of a virtual marketplace where products and services could be explored, compared, and transacted online (Bigliardi et al., 2021). A departure in emphasis from products to the customer experience meant businesses became increasingly interested in customising products and developing personal connections with their customers (Xiang & Fesenmaier, 2017). Inventive online intermediaries such as Expedia emerged and disrupted the traditional travel agency business model. The growth of review sites such as TripAdvisor enabled customers to control their purchases and influence the decision-making of other future consumers (Shiwakoti et al., 2022). For the first time, Visitor Information Centres began to experience a decline in their role as a point of contact as tourists increasingly booked and completed electronic transactions online, expediting the expansion of global markets.

The third stage was described as the 'integration of systems'. This commenced around 2010 and is currently still unfolding. Since 2010, developments such as mobile and wearable technologies, cloud computing, augmented and virtual reality, Global Point System, and increased integration and interoperability of digital systems have expedited interconnectivity between the digital and physical worlds (Mhlanga et al., 2021). Together with collaborative social media platforms and Web 2.0 that enable user-generated content, these developments opened up novel and innovative product development opportunities that have accelerated the global reach of tourism businesses (Dredge et al., 2018). Table 10.1 summarises the three phases of digitalisation in the tourism industry.

Table 10.1 Development of digitalisation in tourism

Phase 1. 1990-2000	Phase 2. 2000-2010	Phase 3. 2010 onwards
Electronic cash registers Financial Software Mobile phones Email Intranet Internet banking Office Software Video conferencing Websites Destination (City) cards	Smartphones Computer graphics software Property management systems Computerised ticketing systems Computerised stock control systems Online booking systems Customer reservation systems Email marketing Customer relationship systems	Augmented reality Virtual reality Mobile Apps Cloud computing and online data storage Wearable technologies Social media Google Analytics Review websites Collaborative online environments Web 2.0 Chatbots and instant advice Peer production, e.g. collaborative platform economy; collaborative commons economy

Source: Dredge et al. (2018)

10.3.2 Digital Tools in Tourism

Digitalisation has forced businesses to change how they engage the market and how tourism services are viewed, consumed, and accessed (Chamboko-Mpotaringa & Tichaawa, 2021). There are several key digital technologies that tourism enterprises are encouraged to explore to increase productivity and create value. These include the use of Artificial intelligence (AI). It is defined as intelligent automation and can be used to support the concept of smart tourism (Tussyadiah, 2020). AI technology can personalise and customise services when requested. This technology can make tourists' travel easier and more comfortable (Organisation for Economic Cooperation and Development (OECD), 2020). Kindzule-Millere and Zeverte-Rivza (2022) note that robots are being created and used in the tourism industry to provide different technical capabilities and perform repetitive, dirty, and dangerous tasks as AI grows. During COVID-19, robots became valuable tools for physical distancing between hosts and guests (Seyitoglu & Ivanov, 2021).

The other tool is Big Data Analytics. This is a digital tool for collecting and analysing bulk data (Shereni & Chambwe, 2020). Data handling is one of the biggest challenges in today's rapidly digitalising business world (Fernindez-Rovira et al., 2021). With the world digitalising, customers and businesses generate large amounts of data (OECD, 2020). Using these data is viewed as a strategy for building customer loyalty and satisfaction since it is generated by customers themselves (Mhlanga & Ndhlovu, 2023b). Using this data enables enterprises to forecast demand, consumers' behaviour, and purchasing inclinations and provides an opportunity for personalisation. Therefore, this becomes an essential aspect of tourism marketing (Fraccastoro et al., 2021). According to Fernindez-Rovira et al. (2021), when the information from big data analyses is appropriately used, enterprises can develop new business models and increase their efficiency.

The Internet of Things (IoT) is another tool with the digitalisation wave. The IoT enables the interoperability of sensors, data, and automation to provide real-time data and information (Ndhlovu & Mhlanga, 2023). The Internet is used for tourism marketing and management to improve the customer experience (Dredge et al., 2018). The Internet enables more urban proficiency through a data-rich tourism industry (Kindzule-Millere & Zeverte-Rivza, 2022). It also supports smart tourism since it improves operational and resource efficiency while reducing environmental impact (OECD, 2020). The Internet works together with other technologies.

Another tool which emerged with the advent of digitalisation is Blockchain technology. This technology is based on a complex digital procedure that aggregates and organises data into blocks and afterwards brings together these blocks using cryptography (Filimonau & Naumova, 2020). For enterprises in the tourism industry, Blockchain technology can be used to manage sales, improve overall management, monitor finance and administration transactions, and improve communication with external stakeholders (Marx et al., 2021). With the cooperation of all stakeholders, Blockchain has the potential to stimulate sustainable tourism (Tyan et al., 2021).

According to Kindzule-Millere and Zeverte-Rivza (2022), Blockchain technology can inspire entrepreneurship and innovation. It enables peer-to-peer entrepreneurship by optimisation of shared economy applications (Busulwa et al., 2022). Blockchain enables digital currencies, optimised product circulation and refined trading models, and digital/crypto-currencies (Filimonau & Naumova, 2020).

Mobile technology/cloud computing is another technology that has emerged under 4IR. Mobile devices are increasingly used to satisfy travel goals, including access to real-time destination information, online booking, and online payments. On the other hand, cloud technology enables enterprises to advertise and offer products and services from anywhere in the world with an internet connection (OECD, 2020). TripAdvisor revealed that about 45 per cent of people used their smartphones in their travelling process (Raga, 2020). Labanauskaite et al. (2020) observed that mobile devices with various applications have become essential in people's daily lives, saving time, being more flexible, and communicating with enterprises and peers through social media. According to Kindzule-Millere and Zeverte-Rivza (2022), mobile phones are now playing the role of travel agencies, tour guides, tour operators, travel maps, and the best restaurant locator. Some app notifications enable enterprises to communicate with potential customers in real time to remind them or suggest services and products.

Augmented reality/virtual reality (AR/VR) are other digital tools quickly gaining prominence in tourism. VR is a tool that creates a more immersive interface, while AR systems exhibit virtual objects in the real world. By description, AR denotes the addition of digital information to the real environment to provide customers with an insight into the world in front of them without actually being there (Ndhlovu & Dube, 2023b). This can be used to explore historical informative, and adventurous experiences or wholly new and imaginary reality. It can comprise text and images that increase the experience of customers (Chamboko-Mpotaringa & Tichaawa, 2021). Wei et al. (2019) posit that VR affords tourists to practically virtually immerse, orient, and interact with different senses. A digital environment allows tourists to explore new and specific places and destinations they would like to visit. AR and VR are indispensable for tourism enterprises' marketing strategies and income development (Akhtar et al., 2021).

Social media has also gained popularity as an effective digital information and engagement tool in recent years. In tourism, the tool is now widely used for creating brand awareness for regions, enterprises, destinations, and attractions and developing relationships with tourists before, during, and after travel (Hysa et al., 2021). It is argued that the advent of social media has led to a paradigm shift in communication, with tourists often choosing their destination based on the information acquired on social media (Ali et al., 2022). As a result, tourism enterprises need to actively engage platforms to keep pace with the kind of information which customers share on social media.

10.3.3 Factors Affecting Digitalisation

Numerous factors affect the digitalisation process within enterprises. These factors include Organisational culture and technological transition; Infrastructural challenges and barriers; Financial challenges and barriers; and Training and Education (Fraccastoro et al., 2021; Kindzule-Millere & Zeverte-Rivza, 2022; Ndhlovu & Dube, 2023; Tyan et al., 2021). Figure 10.1 summarises the factors that affect digitalisation in tourism enterprises.

Organisational Culture and Technological Transition

Every enterprise operates within a specific culture. This culture is a result of differences, which range from social, socio-economic, and political factors, as well as available capital (physical, financial, human, natural, and social) (Ndhlovu & Dube, 2023; Tyan et al., 2021). Organisational culture refers to an organisation's values, beliefs, and norms (Tidd & Bessant, 2018). Digitalisation often requires organisations to change or adjust their cultures to adopt and implement innovations. Organisational culture can hamper the digitalisation process. Alawamleh et al. (2022) found that even where the organisation has the resources and skills to digitalise, culture can exert pressure, much to the frustration of the digitalisation process. In Jordan, culture was found to be the primary barrier to an open technological transition (Alawamleh et al., 2022). It is, therefore, recommended that for enterprises to overcome the problematic role of culture in the digitalisation process,

Fig. 10.1 Factors which affect digitalisation. (Source: Authors, Developed from various sources)

culture must be pursued in a manner that endorses open innovation practices and strategies and not just in the conservative way of rivalry (Cichosz et al., 2020).

According to Maksimenko et al. (2021), an understanding of organisational culture can help the management of an organisation to identify available proficiencies and deficiencies that necessitate whether digitalisation should be gradual or radical. However, evidence shows that culture has obstructed digitalisation in developing countries where enterprises such as restaurants face resource challenges (Ndhlovu & Dube, 2023). Most enterprises in these countries sustain some inflexible cultures as a survival strategy. Enterprises that deploy this strategy defend their operational systems against new or different behaviour, structure, and cognitive approaches to reinforce the status quo. Even where digitalisation is required, such enterprises resist it because it can threaten the meagre resources (Cichosz et al., 2020). In Ghana, it was found that restaurants that did not have unyielding cultures were more willing to adopt digital tools.

In contrast, those with inflexible cultures preferred to keep traditional technology (Kumaning & Godfred, 2019). In Malaysia, Ishak et al. (2021) found that in defence of individual cultures, many restaurants preferred to develop individual technological solutions based on their metrics and methodologies rather than rapidly digitalise in a manner that can threaten existing cultures. This situation was also reported in Jordan, where Alawamleh et al. (2022) found that cultural change resistance resulted in the advent of a range of digital transitions that are often superficial, weak, unappealing to customers, and lack revolutionary potential (Alawamleh et al., 2022).

Financial Challenges and Barriers

In countries such as Ghana (Kumaning & Godfred, 2019), South Africa (Nkosana et al., 2016), Malaysia (Ishak et al., 2021), and Brazil (Khan, 2022), it was found that financial challenges were the key obstacle to digitalisation. It is argued that most challenges enterprises face regarding the digital transition are linked to their financial status. For instance, in South Africa, Nkosana et al. (2016) found that some enterprises, such as restaurants, avoided or postponed technological transitions as a strategy to save their meagre finances. In Asia (Khan, 2022) and Africa (Asefa et al., 2020), financial challenges, ensured by high operational expenses, translated into technological transition and digitalisation challenges. Economic challenges, therefore, compromise the capacity of these enterprises to digitalise.

In most developing countries, some hospitality and tourism enterprises, like restaurants and guesthouses, fail to approach financial institutions for financial support due to inadequate requirements (Bhorat et al., 2018). Most financial institutions in developing countries are much more conformist and traditional and willing to support small enterprises only when the prospect of success is inevitable. For this reason, many small tourism enterprises fail to access the funds they need to digitalise and improve their operations. According to Ndhlovu and Dube (2023), financial challenges for small businesses worsened during COVID-19, in which even financial institutions became stricter regarding money lending.

Social Barriers to Digitalisation

Social barriers related to digitalisation's perceived implications on people's lives and livelihoods, especially employees, have also emerged. The major concern is that technology integration and digitalisation could replace humans with machines, thus, resulting in unemployment (Mhlanga et al., 2021). This view is supported by Cortis et al. (2021), who posit that technological transition will exacerbate unemployment in the short term, irrespective of its long-term beneficial effects. Robertson (2018) also avers that the technological revolution will not destroy work; it will destroy jobs. As a result, in countries with high unemployment rates, governments are very cautious about supporting the digital transition, fearing that this could shatter livelihoods and plunge their populations into poverty and unemployment. In this view, radical digitalisation is unlikely to gain adequate support in these countries, particularly as it threatens to hold back ongoing efforts to create and protect jobs.

Infrastructural Challenges and Barriers

Infrastructure is a challenge for both small and large enterprises. The problem is, however, more pronounced in developing countries (Bhorat et al., 2018; Kamaruzaman et al., 2019). In Brazil, it was found that even though digitalisation was highly esteemed by owners of small businesses such as restaurants, infrastructural constraints were still their main challenge (Crumo, 2022). Amlin et al. (2018) found that infrastructural challenges prevented guesthouses and restaurants from fully digitalising and benefitting from social media technologies in India. In South Africa, enterprises' main focus was on acquiring and developing basic infrastructure rather than acquiring digital tools (Bhorat et al., 2018). Rocha et al. (2016) found that globally, telecommunication infrastructure was concentrated in urban areas, thus, increasing the 'digital divide' between urban and rural enterprises. This has deprived enterprises in remote areas that might be seeking to digitalise (Cirera et al., 2022). Nkosana et al. (2016) aver that reducing the digital divide between urban and rural areas could improve the profitability of the tourism industry and, thus, enable it to contribute to the economy.

In helping enterprises to digitalise rapidly, countries such as Brazil, Thailand, and Chile are rolling out digital infrastructure so enterprises such as restaurants can affordably go digital. As part of its 4IR policies, Thailand is promoting e-commerce platforms and social media applications providing digital infrastructure, such as the nationwide broadband network, to encourage adoption by enterprises (Amornkitvikai et al., 2022; Kohpaiboon, 2020). Brazil and Chile also provide grants to small businesses to ramp up digitalisation (Cirera et al., 2022). This differentiates the trends of technology adoption within and across countries.

Training and Education

Digitalisation requires the presence of adequate technical skills and training. A technical skills gap has been identified by scholars (Alahakoon & Jehan, 2020; Ndhlovu & Dube, 2023). However, the tourism and hospitality industry has recently experienced an exodus of highly skilled personnel, especially following the COVID-19 pandemic (International Labour Office (ILO), 2022). The strict safety requirements such as national and regional lockdowns; restrictions on the movement of people and goods—both national and international; social distancing and cancellation of gatherings; adoption of self-isolation; and mandatory quarantines enormously affected the tourism industry, which thrives on the movement and gathering of people. In Malaysia, Ishak et al. (2021) report that an enormous percentage of skilled and experienced staff switched to other industries, away from tourism, following the pandemic. The exodus of skilled personnel from the industry leaves enterprises with inadequate skilled and trained staff to rapidly implement digitalisation. Successful digitalisation requires a well-trained and well-qualified workforce. Therefore, the absence or inadequate skilled personnel in tourism enterprises also represents another obstacle that could hamper the digitalisation process.

10.3.4 Toward Inclusive Digitalisation in Global Tourism Enterprises

There is no 'one-size fits all' strategy for ensuring digitalisation for both enterprises and employees. As a result, context-specific approaches are needed. This chapter, however, outlines some generic strategies that can be adjusted and improved as part of an effort to encourage inclusive digitalisation. The aim is to help identify appropriate and recommend policy initiatives and actions. In this section, the initiatives for promoting inclusive digitalisation are framed around four major recommendations: Network Development; Education and Training; Financial Support; and Research and Development.

Network Capacity Building

Hospitality and tourism enterprises need to be supported to capitalise on the various networks they already have while working towards additional ones (Troshani et al., 2018). Networks enable the sharing of information and, sometimes, the sharing of resources (Ndhlovu & Dube, 2023). This sharing can empower enterprises to eventually engage in substantial digitalisation, which does not diminish its resources (financial, physical, human, and social). For capacity-building networks to be more effective and support digitalisation commitments, they need to include all stakeholders beyond tourism, including customers (of all demographic characteristics), technology companies, and other complementary actors so as to ensure that digitalisation is as inclusive as possible.

At the individual business level, tourism enterprises should be creative in terms of acquiring resources to innovate and digitalise without compromising the business goals. Network capacity building must expand all forms of partnership or cooperation within the tourism industry and along the value chain (Nkosana et al., 2016). Capacity-building efforts should seek to build platforms that connect enterprises in the industry. Peer-to-peer mentoring, training initiatives, and networking opportunities can also strengthen the digitalisation wave. Enterprise-to-enterprise platforms should be linked with larger online advertising companies to extend tourism products and services to more destinations and larger markets. National and local governments should play an important role in these initiatives, given that some of the infrastructure needed requires governmental intervention.

Education and Training

Inclusive digitalisation requires the rolling out of Interactive education. Enterprises should not be reduced to mere receivers of digital education but should be allowed to be active participators in the education process. Dredge et al. (2018) noted that the current educational support provided by public administrations and professional associations to small businesses strongly focuses on improving general knowledge of digitalisation using webinars, newsletters, and information sessions, for instance. This type of education is passive and could confound and confuse enterprise owners, especially those in older age groups, thus preserving the perceived challenges of digitalisation. It is, therefore, important to ensure that education approaches are context-specific and to provide more interactive sessions that include an in-depth discussion of, for instance, the diverse stages of digitalisation and approaches to moving from traditional business to e-business, e-commerce, and smart tourism. Interactive education should also enable business owners and their employees to explore the various digital tools that can be leveraged to improve products and services using novel tools such as big data analytics, mobile applications, interconnectivity, IoT, and VR.

Interactive education could also help enterprises to understand the cost versus return on investment of digitalisation. The benefits of digitalisation should be presented to enable tourism enterprises to make thought-out decisions based on not only monetary gains but also possible long-term immaterial paybacks such as improved effectiveness, control, partnership, and data quality. Through interactive discussions, enterprises may be more enthusiastic about investing in digitalisation activities.

Like education, training needs to be differentiated and targeted to different groups of tourism enterprises. These enterprises are already in various stages of digitalisation, with some still moving into e-business and others transitioning to e-commerce, while others are already highly digitalised and experimenting with new technologies (Rondović et al., 2019; Santos et al., 2020). It was found that most small businesses, including small tourism businesses, needed more skills for digital marketing, social media, web development, cyber-security, and data analytics (European Commission, 2018). Dredge et al. (2018) observed that across Europe, public administrations and professional associations seemed to focus on basic training opportunities for tourism enterprises, such as digital skills for online promotion (the development of websites, social networking) and sales (online booking and reservation systems and customer management. Training programmes for inclusive digitalisation should, however, take into account challenges such as upskilling in cyber-security, data analytics, and the needs of different categories of stakeholders. Dredge et al. (2018) observed that small tourism businesses found it more convenient and valuable to learn directly from relevant industry experts and peers rather than rely on public administrations and professional associations. An effective and efficient training agenda that can generate inclusive digitalisation, therefore, can balance between formal training and semi-formal/informal training through peer-to-peer programmes.

Financial Support

Some countries have policies in place to support small businesses (Amornkitvikai et al., 2022; Kohpaiboon, 2020). However, this kind of support does not exist in most developing countries. In countries such as Ghana, South Africa, Malaysia, and Brazil, it was found that the lack of financial support was the key obstacle to digitalisation (Ishak et al., 2021; Kumaning & Godfred, 2019; Khan, 2022). Thus, financial support could help with training and other related costs to improve digitalisation in enterprises. With most enterprises hesitant to use the meagre resources to acquire and implement new technologies (Nkosana et al., 2016), financial support could provide some relief and motivate these enterprises to integrate digital tools into their operations. For tourism enterprises located in remote areas where digital infrastructure is inadequate, financial support can be used to develop infrastructure to reduce the costs of broadband and mobile services.

Research and Development

Above and beyond direct skill training programmes and financial support targeted SMEs, there is a need for research to identify the specific digital skills that are important for tourism enterprises. Research is also needed to explore ways of tailoring digitalisation to the operational needs of individual enterprises. Therefore, research, development, and innovation are key components of digitalisation and its inclusiveness and sustainability in tourism. In the tourism industry, the benefits of digitalisation exist on a continuum which scales from the individual enterprise to the collective interactions generated through inclusion in digital supply chains through to global value networks (Dredge et al., 2018). Digitalisation should be considered a step for integrating small tourism businesses into broader economic value chains. Research and innovation are also important for finding solutions to specific tourism-related societal challenges, including over-tourism and sustainability in tourism.

10.4 Conclusion

This chapter explored the strategies which may be used to promote digitalisation in global tourism enterprises. Focus has been placed on small enterprises since they often experience resource challenges compromising the potential to digitalise. The chapter shows that factors such as organisational culture, financial factors, infrastructural challenges, social barriers, and education and training are the main factors that affect digitalisation in both developed and developing countries. As a result, to promote digitalisation; and make it as inclusive as possible, addressing these factors should be the starting point. With most small tourism enterprises, particularly in developing countries, being relatively traditionalist in their business approaches due to limited resources, financial support can promote a change of attitude towards embracing digital tools. Financial support should also be supported by interactive education and training to enable enterprise owners and employees to appreciate digitalisation opportunities, particularly in terms of more efficient business practices and market expansion, to implement digital solutions to increase their attractiveness, grow, and expand their networks. Future research should explore how large tourism companies can be networked with their smaller cousins so that the two categories can benefit from their digital prowess of each other.

References

Akhtar, N., Khan, N., Khan, M. M., Ashraf, S., Hashmi, M. S., Khan, M. M., & Hishan, S. S. (2021). Post-COVID-19 tourism: Will digital tourism replace mass tourism? *Sustainability, 13*(5352), 1–18.

Alahakoon, M. U. I., & Jehan, S. N. (2020). Efficiency of public service delivery—A post-ICT deployment analysis. *Economies, 8*(4), 97.

Alawamleh, M., Al-Hussaini, M., & Ismail, L. B. (2022). Open innovation in the food industry: Trends and barriers: A case of the Jordanian food industry. *Journal of Global Entrepreneurship Research, 12*, 279. https://doi.org/10.1007/s40497-022-00312-6

Ali, I., Azhar, N., Tariq, M., & Khan, S. A. (2022). Role of rise in social media on travel and tourism industry. *International Journal of Research and Analytical Reviews, 9*(3), 1150–1155.

Amornkitvikai, Y., Tham, S. Y., Harvie, C., & Buachoom, W. W. (2022). Barriers and factors affecting the E-commerce sustainability of Thai micro-, small- and medium-sized enterprises (MSMEs). *Sustainability, 14*, 8476.

Amlin, D., Sudhahar, C., & Varghese, J. (2018). Digital technology integration for small restaurant business in India. *International Journal of Civil Engineering and Technology, 9*(6), 1593–1601.

Asefa, A., Qanche, Q., Hailemariam, S., Dhuguma, T., & Nigussie, T. (2020). Risk perception towards COVID-19 and its associated factors among waiters in selected towns of southwest ethiopia. *Risk Management and Healthcare Policy, 13*, 2601.

Bhorat, H., Asmal, Z., Lilenstein, K., & Van der Zee, K. (2018). *SMMES in South Africa: Understanding the constraint* (Working papers). DPRU, University of Cape Town.

Bigliardi, B., Ferraro, G., Filippelli, S., & Galati, F. (2021). The past, present and future of open innovation. *European Journal of Innovation Management, 24*(4), 1130–1161.

Busulwa, R., Pickering, M., & Mao, I. (2022). Digital transformation and hospitality management competencies: Toward an integrative framework. *International Journal of Hospitality Management, 102*(103132), 1–16.
Chamboko-Mpotaringa, M., & Tichaawa, T. M. (2021). Tourism digital marketing tools and views on future trends: A systematic review of literature. *African Journal of Hospitality, Tourism and Leisure, 10*(1), 712–726.
Cichosz, M., Wallenburg, C. M., & Knemeyer, A. M. (2020). Digital transformation at logistics service providers: Barriers, success factors and leading practices. *The International Journal of Logistics Management, 31*(2), 209–238.
Cirera, X., Comin, D., & Cruz, M. (2022). *Bridging the technological divide: Technology adoption by firms in developing countries*. World Bank Group.
Cortis, N., Smyth, C., Valentine, K., Breckenridge, J., & Cullen, P. (2021). Adapting service delivery during COVID-19: Experiences of domestic violence practitioners. *The British Journal of Social Work, 51*(5), 1779–1798.
Crumo, C. A. (2022). Delivering meanings: The 'Appification' of formal restaurants in Brazil amid the coronavirus pandemic. *Food, Culture and Society*. https://doi.org/10.1080/15528014.2022.2143160
Dredge, D., Phi, G., Mahadevan, R., Meehan, E., & Popescu, E. S. (2018). *Digitalisation in tourism: In-depth analysis of challenges and opportunities. Low value procedure GRO-SME-17-C-091-A for Executive Agency for Small and Medium-sized Enterprises (EASME) Virtual Tourism Observatory*. Aalborg University.
European Commission. (2018). *Upskilling Europe's small businesses for the digital age*. Upskilling Europe's small businesses for the digital age | Shaping Europe's digital future (europa.eu). Accessed 14 June 2023.
Fernindez-Rovira, C., Valdes, J. I., Mollevk, G., & Nicolas-Sans, R. (2021). The digital transformation of business. Towards the datafication of the relationship with customers. *Technological Forecasting and Social Change, 162*, 1–12.
Filimonau, V., & Naumova, E. (2020). The blockchain technology and the scope of its application in hospitality operations. *International Journal of Hospitality Management, 87*, 1–8.
Fraccastoro, S., Gabrielsson, M., & Pullins, E. B. (2021). The integrated use of social media, digital, and traditional communication tools in the B2B sales process of international SMEs. *International Business Review, 30*(4), 101776, 1–15.
Gokkaya, S. (2020). Chapter: 11: Digital transformation and marketing in tourism industry. In *Selected academic studies from Turkish tourism sector* (pp. 149–164). Peter Lang GmbH.
Hassan, S. C. (2023). Innovative marketing strategies during tourism recovery in the post-COVID-19 period in Nigeria. In P. M. A. S. A. H. J. Kennell (Ed.), *Crisis management, destination recovery and sustainability* (pp. 67–76). Routledge.
Hysa, B., Karasek, A., & Zdonek, I. (2021). Social media usage by different generations as a tool for sustainable tourism Marketing in Society 5.0 idea. *Sustainability, 13*, 1018, 1–27.
ILO. (2022). World employment and social outlook: Trends 2022. ILO: Geneva.
Ishak, F. A. C., Mazlan, N., Lokman, N. A., Karim, M. S. A., & Mohamad, S. F. (2021). Exploring the issues and challenges in managing themed restaurant in Klang Valley, Malaysia. *International Journal of Academic Research in Business and Social Sciences, 11*(10), 1121–1135.
Kamaruzaman, F. M., Hamid, R., Mutalib, A. A., & Rasul, M. S. (2019). Conceptual framework for the development of 4IR skills for engineering graduates. *Global Journal of Engineering Education, 21*(1), 54–61.
Khan, M. A. (2022). Barriers constraining the growth of and potential solutions for emerging entrepreneurial SMEs. *Asia Pacific Journal of Innovation and Entrepreneurship*. https://doi.org/10.1108/APJIE-01-2022-0002
Kindzule-Millere, I., & Zeverte-Rivza, S. (2022). Digital transformation in tourism: Opportunities and challenges. In *Proceedings of the 2022 international conference "Economic science for rural development" No 56 Jelgava, LLU ESAF, 11–13 May 2022* (pp. 476–486). https://doi.org/10.22616/ESRD.2022.56.047478

Kohpaiboon, A. (2020). *Industry 4.0 policies in Thailand* (Economic Working Paper No. 2020–02). ISEAS Yosof Ishak Institute.

Kumaning, R. G., & Godfred, A. (2019). Challenges of E-commerce adoption in the Ghanaian hotel industry. *European Journal of Business and Management, 11*(27), 126–134.

Kusuma, H., Muafi, M., Aji, M. H., & Pamungkas, S. (2020). Information and communication technology adoption in small- and medium-sized enterprises: Demographic characteristics. *Journal of Asian Finance, Economics and Business, 7*(10), 969–980.

Labanauskaite, D., Fioreb, M., & Stasysa, R. (2020). Use of E-marketing tools as communication Management in the Tourism Industry. *Tourism Management Perspectives, 34*(100652), 1–8.

Marx, S., Flynn, S., & Kylanen, M. (2021). Digital transformation in tourism: Modes for continuing professional development in a virtual community of practice. *Project Leadership and Society, 2*(100034), 1–11.

Maksimenko, I., Vashko, T., & Zdrestova-Zakharenkova, S. (2021). Digital Transformation and its challenges to the strategic management system. In SHS web of conferences (Vol. 106). EDP Sciences.

Mhlanga, D & Ndhlovu, E (2023a). 'The Fourth'. In Mhlanga, D and Ndhlovu, D. (Eds.). The fourth industrial revolution in Africa: Exploring the development implications of smart technologies in Africa. Springer.

Mhlanga, D & Ndhlovu, E. (2023b). Making sense of the Fourth industrial revolution: An overview of the potential impacts on Africa. In D. Mhlanga & D. Ndhlovu (Eds.), *The fourth industrial revolution in Africa: Exploring the development implications of smart technologies in Africa*. Springer.

Mhlanga, D., Ndhlovu, E., & Hofisi, C. (2021). Assessment of the 4IR challenges of radical innovation in service delivery in Africa. *Journal of Public Administration, 56*(4.1), 1002–1017.

Ndhlovu, E., & Dube, K. (2023). Challenges of radical technological transition in the restaurant industry within developing countries. *African Journal of Hospitality, Tourism and Leisure, 12*(1), 156–170.

Ndhlovu, E., & Mhlanga, D. (2023). Smart technologies, climate change, and smallholder farming in Zimbabwe. In D. Mhlanga & D. Ndhlovu (Eds.), *The fourth industrial revolution in Africa: Exploring the development implications of smart technologies in Africa*. Springer.

Nkosana, T., Skinner, J., & Goodier, C. (2016). Challenges of ICT adoption and utilisation in small rural restaurants: A case study of the KwaZulu-Natal midlands. *African Journal of Hospitality, Tourism and Leisure, 5*(4), 1–11.

OECD (Organisation for Economic Co-operation and Development). (2020). *Tourism trends and policies 2020*. Available on: https://www.oecd-ilibrary.org/sites/f528d444-en/index.html?itemId=/content/component/f528d444-en. Accessed 13 June 2023.

Opute, A. P., Irene, B. O., & Iwu, C. G. (2020). Tourism service and digital technologies: A value creation perspective. *African Journal of Hospitality, Tourism and Leisure, 9*(2), 1–18.

Raga, J. (2020). *Tourism informatics* (p. 273). Society Publishing. APA 7th Edition (American Psychological Assoc.) MLA 8th Edition (Modern Language Assoc.).

Robertson, S. L. (2018). The role of technology in service delivery for people with disabilities in rural communities. In D. A. Harley, N. A. Ysasi, M. L. Bishop, & A. R. Fleming (Eds.), *Disability and vocational rehabilitation in rural settings* (pp. 59–76). Springer.

Rondović, B., Djuričković, T., & Kašćelan, L. (2019). Drivers of E-business diffusion in tourism: A decision tree approach. *Journal of Theoretical and Applied Electronic Commerce Research, 14*(1), 30–50.

Rocha, Á., Correia, A. M., Adeli, H., Reis, L. P., & Teixeira, M. M. (2016). New advances in information systems and technologies. Springer.

Santos, F., Lunardi, G. L., Maia, C. R., & Anãnã, E. S. (2020). Factors that influence consumers' participation in electronic tourism. *Revista Brasileira de Pesquisa em Turismo, São Paulo, 14*(2), 139–155.

Schallmo, D., Williams, C. A., & Boardman, L. (2017). Digital transformation of business models – Best practice, enablers, and roadmap. *International Journal of Innovation Management, 21*(8), 1–17.
Seyitoglu, F., & Ivanov, S. (2021). Service robots as a tool for physical distancing in tourism. *Current Issues in Tourism, 24*(12), 1631–1634.
Shereni, N. C., & Chambwe, M. (2020). Hospitality big data analytics in developing countries. *Journal of Quality Assurance in Hospitality & Tourism, 21*(3), 361–369.
Shiwakoti, N., Hu, Q., Pang, M. K., Cheung, T. M., Xu, Z., & Jiang, H. (2022). Passengers' perceptions and satisfaction with digital technology adopted by airlines during COVID-19 pandemic. *Future Transportation, 2*, 988–1009.
Skoczylas, M. J. (2019). Congruence in research question, design, and analysis: A tutorial on the measurement of change in clinical speech and language research. *Language, Speech, and Hearing Services in Schools, 50*(2), 167–178.
Tidd, J., & Bessant, J. (2018). *Managing innovation: Integrating technological, market and organizational change* (Enhanced eText, 6th Edition) (VitalSource Bookshelf version). Retrieved from vbk://9781119379416.
Troshani, I., Janssen, M., Lymer, A., & Lee, D. (2018). Digital transformation of business-to-government reporting: An institutional work perspective. *International Journal of Accounting Information Systems, 31*, 17–36.
Tussyadiah, I. (2020). A review of research into automation in tourism: Launching the annals of tourism research curated collection on artificial intelligence and robotics in tourism. *Annals of Tourism Research, 81*(102883), 1–13.
Tyan, I., Yague, M. I., & Guevara-Plaza, A. (2021). Blockchain technology's potential for sustainable tourism. In W. Worndl et al. (Eds.), *Information and communication technologies in tourism* (pp. 17–29). Springer.
Vaismoradi, M., & Snelgrove, S. (2019). Theme in qualitative content analysis and thematic analysis. *Forum: Qualitative Social Research, 20*(3), 1–14.
van Dyk, R., & Van Belle, P. (2019). Factors influencing the intended adoption of digital transformation: A south African case study. *Computer Science and Information Systems, 18*, 519–528.
Wei, W., Qi, R., & Zhang, L. (2019). Effects of virtual reality on theme park visitors' experience and behaviors: presence perspective. *Tourism Management, 71*, 282–293.
Xiang, Z., & Fesenmaier, D. (2017). Big data analytics, tourism design and smart tourism. In *Analytics in smart tourism design: Concepts and methods* (pp. 299–307). Springer.
Zeqiri, A., Dahmani, M., & Youssef, A. B. (2020). Digitalization of the tourism industry: What are the impacts of the new wave of technologies. *Balkan Economic Review, 2*, 63–82.

Chapter 11
Challenges of Technology Integration in the Tourism Industry in Africa

Emmanuel Ndhlovu and Tshenolo Montsiemang

11.1 Introduction and Background

The tourism and hospitality industry is significantly impacted by technological improvements. The importance of technology integration in the industry for achieving success is apparent based on an extensive assessment of the existing literature. According to Law et al. (2014), the incorporation of information communication technologies (ICTs) assumes a significant role within the domains of tourism, travel, and hospitality industries. Ndhlovu and Dube (2023a) affirm that the integration of information and communication technology (ICT) within the tourism and hospitality industry has emerged as a crucial factor for achieving economic success. This integration holds significant importance for a multitude of reasons. First, it facilitates the improvement of communication and collaboration among tourism groups and service suppliers. Furthermore, it enables the acquisition and dissemination of market intelligence, a critical component for the advancement and provision of efficient tourism services (Alford & Jones, 2020). Chen et al. (2021) state that the integration of technology in tourism enterprises boosts the experiences and happiness of customers through the provision of personalised and seamless services. Moreover, the incorporation of technology facilitates the effective administration of several operational aspects, such as the management of reservations, bookings, and inventory control (Ozdemir et al., 2023). Moradi et al. (2023) asserted that the integration of technology within the tourist and hospitality sector plays a crucial role in the promotion and marketing of various tourism destinations, attractions, and accommodations. Few studies focus on technology adoption challenges with a focus on developing regions such as Africa. This chapter seeks to close this gap by exploring technology challenges in Africa.

E. Ndhlovu (✉) · T. Montsiemang
Vaal University of Technology, Vanderbijlpark, South Africa
e-mail: tshenolom@vut.ac.za

E. Ndhlovu et al. (eds.), *Tourism and Hospitality for Sustainable Development*,
https://doi.org/10.1007/978-3-031-63069-9_11

There are few studies which focus on technology adoption and integration in tourism in Africa. Ade-Ibijola and Okonkwo (2023) and Cloete (2021) explored artificial intelligence on the continent but with no focus on the tourism industry. Amankwah-Amoah (2019) studied technology and sustainability and did not explore adoption and integration challenges in tourism. Nyagadza et al. (2022) studied technology adoption but focused on the events sub-sector, and therefore, did not provide a general overview of the tourism industry in Africa. Chihwai et al. (2023) explored the importance of digital technology in destination decisions by tourists but did not explore the challenges of technology adoption. Ndhlovu and Dube (2023b) proposed digital tools as the potential resilience strategy after the COVID-19 pandemic but did not expand on the challenges of technology access and adoption. As a result, our understanding of the technology adoption and integration challenges in the tourism industry remains limited. This has huge practical and policy implications.

This chapter, therefore, (i) explores the challenges of digitalisation and technological integration in tourism in Africa and (ii) proposes intervention strategies. The study has much potential in terms of adding to the existing scholarship on the topic and also to policy interventions in the sector. The study could provoke action by the relevant stakeholders to begin imaging technology adoption and integration in tourism in a manner that generates sustainability and profitability.

The chapter is organised as follows: The next section outlines the research methodology for the study. The chapter then details the challenges of technology adoption in the tourism industry in Africa. This is followed by a discussion on how the challenges identified can be overcome. Thereafter, future perspectives on technology adoption and integration are presented. Lastly, conclusions are drawn from the discussion.

11.2 Methods and Materials

The chapter follows a conceptual research approach. It is based on the analysis of existing debates to highlight some of the challenges that can emerge from technological integration. Conceptual research does not involve practical experiments (Govender et al., 2022). It is related to abstract concepts or ideas. Conceptual research tries to resolve problems by thinking about them. The approach uses the brain and not the hands to solve problems (Van der Waldt, 2020). The study used no systematic method to identify the sources of the debates used in the study. Instead, the study consulted the most recent sources on the use of technology in tourism in Africa. The sources consulted included newspapers, academic publications, and policy documents. These sources were then used to construct the conceptual arguments made in this study. The authors acknowledge that the conceptual nature of this chapter renders it to strong criticism because conceptual research makes conclusions that are not fact-based and therefore, not necessarily reliable. It is, however, hoped that future research will problematise the arguments made here using empirical evidence.

11.3 Presentation and Discussion of Findings

11.3.1 Challenges in Implementing Technology in Tourism

Despite the manifold advantages associated with the integration of technology in the tourism and hospitality industry, there exist certain obstacles that impede its successful implementation (Khatter, 2023). One of the foremost issues lies in the financial implications entailed in the implementation and upkeep of technological infrastructure. The financial burden associated with procuring and enhancing technology systems and software can provide a significant obstacle for numerous small- and medium-sized enterprises operating in the tourism and hospitality sector. Consequently, these enterprises may encounter difficulties in adapting to the swift advancements in technology, thereby impeding their capacity to deliver the same calibre of technical innovation and services as their larger counterparts as emphasised in the study by Otemoyolo (2021). In support of this argument, Mapeshoane and Pather (2016) highlight that one other obstacle pertains to the reluctance to embrace change and the insufficient level of technological proficiency exhibited by personnel. There exists a potential reluctance among employees in the tourism and hospitality industry to embrace novel technologies, primarily stemming from concerns about job displacement and limited comprehension of the appropriate utilisation of technology within their professional domain (Rodriguez Sanchez et al., 2020).

Moreover, the expeditious rate of technology progress presents an inherent problem. Afolayan and De la Harpe (2020) mentioned that in light of the on-going emergence of novel technologies, organisations often encounter challenges in staying abreast of these advancements and discerning the technologies that hold the greatest pertinence and utility for their operational needs. In addition, the integration of technology in the tourism and hospitality industry is confronted with substantial hurdles pertaining to data security and privacy concerns as divulged by Yallop et al. (2023). The maintenance of customer data security and privacy is of utmost importance, given that any unauthorised access or mishandling of personal information can result in significant repercussions for both the corporate entity and the individual consumer (Madhav & Tyagi, 2022). Carlson and Isaacs (2018) further added that the incorporation of technology within the tourism and hospitality sector presents difficulties associated with the digital divide. This refers to the disparity that exists between individuals who possess access to and proficiency in technology, and those who lack such access and proficiency. The digital gap poses a potential constraint on the scope and efficacy of technology integration endeavours, given that disparate levels of access to technology and digital resources exist among individuals and communities (Parsons et al., 2023). This highlights that the expeditious rate of technical progress presents a formidable obstacle in its own right.

Keeping abreast of the most recent technological advancements can be a significant challenge for enterprises operating within the tourism and hospitality industry. The continuous emergence of new technologies necessitates a persistent commitment of work and resources in order to be knowledgeable and up-to-date on the most recent advancements. Lamm (2021) generalised that the incorporation of technology within the tourism and hospitality industry poses a multitude of obstacles. The challenges encompassed in this context pertain to the difficulties associated with adapting to swift technological advancements, encountering resistance towards change and encountering a lack of technological proficiency among employees, grappling with concerns regarding data security and privacy, addressing the digital divide, and the necessity for on-going updates and resources to remain abreast of the latest technologies. Moreover, the use of technology within the tourism and hospitality industry may provide difficulties associated with the evolving demands of customers (Nam et al., 2021). Buhalis et al. (2023) emphasised that in the contemporary business landscape, there is a growing expectation among customers for experiences that are both seamless and personalised. However, the achievement of such experiences through the integration of technology poses a considerable challenge.

11.3.2 Overcoming Technology Integration Difficulties in Tourism

In order to surmount these hurdles and effectively incorporate technology into the tourism and hospitality sector, a number of tactics can be employed. First and foremost, the tourism and hospitality industry must spend financial resources towards the development and maintenance of technology infrastructure (Jawabreh et al., 2022). By allocating resources towards the acquisition and implementation of technology systems and software, enterprises may establish a robust infrastructure that enables them to deliver services with enhanced efficiency and foster a culture of innovation. Goswami and Daultani (2022), further add that businesses must give precedence to the implementation of training and education initiatives aimed at augmenting the technology literacy of their workforce. Businesses may effectively facilitate the integration of technology and enable their staff to comprehend its advantages in their professional endeavours through the provision of adequate training and assistance. The study conducted by Mzembe et al. (2019) also emphasised that the establishment of collaborations and partnerships between entities in the tourism and hospitality industry and technology firms might serve as a means to address and surmount prevailing obstacles. Through establishing partnerships with technology companies, businesses have the opportunity to leverage the knowledge and resources provided by these entities, facilitating the effective incorporation of technology integration (Ogutu et al., 2023).

The study by Ruel and Njoku (2021) argued that the involvement of industry associations and organisations is of paramount importance in the promotion and facilitation of technological integration. These entities can offer assistance, reinforcement, and access to various tools for enterprises to effectively manage the intricacies associated with the incorporation of technology. In support of this argument,

Hughes et al. (2022) emphasise the significance for enterprises to remain knowledgeable regarding the most recent technological breakthroughs and market trends. By remaining abreast of technological advancements, firms can effectively recognise prospects and feasible remedies for surmounting integration obstacles. Therefore, enterprises in the tourism and hospitality industry must place utmost importance on customer experience and engagement during the process of incorporating technology. Businesses can guarantee that integration activities are motivated by the goal of giving value to customers by prioritising the enhancement of the customer experience through the utilisation of technology (Tom Dieck & Han, 2022).

In summary, the incorporation of technology within the tourism and hospitality sector presents certain obstacles. Nevertheless, businesses can surmount these challenges and capitalise on the integration of technology by employing various strategies. These strategies encompass investing in technology infrastructure, offering comprehensive training and education programmes, cultivating collaborations and partnerships, seeking guidance from industry associations, staying abreast of the latest technological advancements, and placing a premium on enhancing customer experience. By incorporating technology, the tourism and hospitality industry can attain significant financial savings, heightened efficiency, and improved income in the long run (Ivanov & Webster, 2019).

The literature research demonstrates that the incorporation of technology into the tourism and hospitality industry is crucial for achieving success in the contemporary digital world. According to Ozdemir et al. (2023), the incorporation of technology into the tourism and hospitality industry is of paramount importance in the contemporary day, as it plays a pivotal role in optimising corporate processes and promoting customer satisfaction. In general, the literature assessment elucidates that the integration of technology in the tourist and hospitality sector offers a range of prospects as well as obstacles. The potential benefits within this context include expense reduction, heightened efficiency, and augmented revenue generation. However, challenges include various factors such as constraints in infrastructure, resistance towards adopting change, inadequate technology skills and expertise, worries regarding data security, and the need to ensure data privacy (Erol et al., 2002). Moreso, it also underscores the existence of many obstacles that necessitate resolution in order to achieve successful integration.

The existing debates on technology integration in tourism underscore the significance of incorporating technology within the tourism and hospitality industry. It is equally imperative for educators to integrate technology into the curriculum as a means of augmenting students' technological proficiencies and equipping them for forthcoming challenges (Cooper et al., 2023). In addition, the literature study elucidates the profound influence that information and communication technologies have had on the tourism and hospitality industry. Ozdemir et al. (2023) assert that these technologies possess the potential to transform industrial practices, strategies, and structures, resulting in enhanced operational effectiveness, decreased expenditures, and enhanced customer satisfaction. Furthermore, the literature analysis highlights the imperative need to include technology in order to effectively cater to the requirements of contemporary tourists. Contemporary tourists exhibit a growing dependence on technology as a means to facilitate several aspects of their travel endeavours, including trip planning, hotel reservations, information retrieval, and the dissemination of their experiences through social media platforms (Anaya & Lehto, 2020).

In light of the prevailing circumstances, it is apparent that the incorporation of technology into the tourist and hospitality sector is important to ensure its prosperity in the contemporary era of digitalisation. The reviewed debates demonstrate the significant necessity of integrating technology within the tourism and hospitality industry. In the contemporary digital world, the imperative of augmenting client experiences, optimising business operations, and maintaining competitiveness is paramount (Nabila, 2023).

11.3.3 Future Perspectives: Technology and the Evolution of Tourism

The tourism and hospitality industry is undergoing significant transformation following the advent of the 4IR. This results in enhanced operational effectiveness, decreased expenditures, and enhanced customer satisfaction. According to Jawabreh et al. (2022), the integration and utilisation of technology have brought about significant changes in the operational methods, strategic approaches, and organisational frameworks within the tourist and hospitality industry, hence facilitating the emergence of novel business models. Buhalis et al. (2023) mentioned that the tourism and hospitality industry has effectively utilised information and communication technologies to provide tourists with inventive services and tailored experiences. The rise of the notion of smart tourism has been a result of utilising mobile digital connectivity to enhance the intelligence of tourism systems (Lee et al., 2020). Furthermore, the incorporation of technology has also brought about a transformation in the manner in which tourists get information, organise their travel arrangements, and communicate their experiences. In support of this statement, Buhalis et al. (2019) emphasised that in contemporary times, tourists heavily depend on technological advancements to facilitate several aspects of their travel endeavours, including but not limited to trip planning, hotel reservations, information retrieval, and the dissemination of their experiences through social media platforms.

One of the primary obstacles encountered in the incorporation of technology within the tourist and hospitality sector pertains to the constraints imposed by existing infrastructure, which presents a considerable impediment to the successful integration of technology in this business (Nam et al., 2021). Myovella et al. (2020) mentioned that numerous locations, especially those situated in developing nations, particularly within the African continent, may exhibit a deficiency in the requisite infrastructure to facilitate sophisticated technological capabilities, such as dependable internet access and digital payment systems. This phenomenon has the potential to impede the smooth incorporation of technology into various aspects of tourism operations, hence constraining the industry's capacity for expansion and advancement. In addition, research conducted by Fu et al. (2022), shows that the integration of technology in the tourism and hospitality industry encounters the obstacle of reluctance to change. Potential resistance may arise from industry experts who are accustomed to conventional practices and exhibit reluctance towards embracing novel technologies.

Moreover, a significant obstacle to the incorporation of technology within the tourism and hospitality industry is the insufficient proficiency and understanding of technology among professionals in the field. A considerable number of individuals working in the tourist and hospitality sector may lack the requisite expertise and understanding to proficiently employ and incorporate technology within their organisational processes (Mingotto et al., 2021). Insufficient proficiency in technology can pose obstacles to the effective integration of technological solutions and limit the industry's capacity to fully exploit technological progress. A study by Yaqoob et al. (2023) also highlights that the incorporation of technology in the tourism and hospitality industry is confronted by substantial obstacles related to data security and privacy. The sector is confronted with the task of safeguarding the security and privacy of sensitive consumer information due to the growing dependence on technology for its storage and transmission. The difficulty is compounded by the rising occurrence of cyber-attacks and data breaches, which can result in significant repercussions for both enterprises and consumers (Mnyakin, 2023).

In order to tackle these difficulties, industry stakeholders must provide utmost importance to the advancement of digital infrastructure in destinations, with a special focus on emerging African countries and other regions across the globe. It is imperative to prioritise initiatives aimed at enhancing internet connectivity and digital payment systems to effectively incorporate technological advancements into tourism operations. Dredge et al. (2019) mentioned the importance of conducting training and education programmes aimed at augmenting the technological abilities and knowledge of professionals inside the sector. These programmes offer industry professionals the essential tools and experience required for proficiently utilising and incorporating technology within their operational processes. In addition, it is imperative to foster collaboration among various industry stakeholders, governments, and technology suppliers (Garanti et al., 2023), to effectively address the obstacles associated with the integration of technology within the tourist and hospitality sector. Through collaboration, these various stakeholders have the opportunity to collectively devise strategies and solutions aimed at effectively overcoming the obstacles hindering the integration of technology.

Lastly, it is crucial to acknowledge the profound influence of globalisation on the tourism and hospitality sector. The phenomenon of globalisation has resulted in heightened levels of competition and necessitated the expansion of tourist enterprises to accommodate a wide array of clientele. According to the study by Lu et al. (2021), the integration of information technology into tourism-oriented organisations is necessary to enhance sustainability and competitiveness. Moreover, the incorporation of technology has the potential to enhance operational efficiency and decrease operational expenditures for service providers within the tourism sector.

11.4 Conclusion

Navigating Technological Challenges in Tourism

In conclusion, the incorporation of technology into the tourism and hospitality sector is of paramount importance for ensuring its long-term viability and ability to remain competitive within the context of a globalised world, of which Africa is an integral part. Nevertheless, some obstacles must be confronted to effectively manage the trajectory towards the successful integration of technology. Industry experts must cultivate the requisite technological literacy to proficiently harness technology within their operational frameworks. In addition, it is imperative to meticulously address concerns regarding data security and privacy to establish and maintain the trust and confidence of customers.

The present literature review emphasises the significance of incorporating technology into the tourism and hospitality industry. This statement underscores the impact of technology on enhancing the efficacy and potency of tourism services, diminishing expenses, and giving rise to the advent of intelligent tourism. Nevertheless, it is important to realise the inherent difficulties associated with the integration of technology. These problems encompass the limited technological literacy among experts in the industry, as well as apprehensions over data security and privacy. The tourism and hospitality industry should not allow these challenges to hinder its adoption of technology; instead, it should regard them as prospects for advancement and enhancement. To ensure success and sustained expansion in the future, the tourism and hospitality industry must prioritise the integration of technology.

References

Ade-Ibijola, A., & Okonkwo, C. (2023). Artificial intelligence in Africa: Emerging challenges. In D. O. Eke, K. Wakunuma, & S. Akintoye (Eds.), *Responsible AI in Africa. Social and cultural studies of robots and AI*. Palgrave Macmillan. https://doi.org/10.1007/978-3-031-08215-3_5

Afolayan, A., & De la Harpe, A. (2020). The role of evaluation in SMMEs' strategic decision-making on new technology adoption. *Technology Analysis & Strategic Management, 32*(6), 697–710.

Alford, P., & Jones, R. (2020). The lone digital tourism entrepreneur: Knowledge acquisition and collaborative transfer. *Tourism Management, 81*, 104139.

Amankwah-Amoah, J. (2019). Technological revolution, sustainability, and development in Africa: Overview, emerging issues, and challenges. *Sustainable Development, 27*(5), 910–922.

Anaya, G., & Lehto, X. (2020). Traveler-facing technology in the tourism experience: A historical perspective. *Journal of Travel & Tourism Marketing, 37*(3), 317–331.

Buhalis, D., et al. (2019). Technological disruptions in services: Lessons from tourism and hospitality. *Journal of Service Management, 30*(4), 484–506.

Buhalis, D., Leung, D., & Lin, M. (2023). Metaverse as a disruptive technology revolutionising tourism management and marketing. *Tourism Management, 97*, 104724.

Carlson, A., & Isaacs, A. (2018). Technological capital: an alternative to the digital divide. *Journal of Applied Communication Research, 46*(2), 243–265.

Chen, T., Guo, W., Gao, X., & Liang, Z. (2021). AI-based self-service technology in public service delivery: User experience and influencing factors. *Government Information Quarterly, 38*(4), 101520.

Chihwai, P., Ndhlovu, E., & Dube, K. (2023). Exploring trends of South African tourist accommodation occupancy during COVID-19. In K. Dube, I. Mensah, & L. Chapungu (Eds.), *COVID-19, tourist destinations and prospects for recovery*. Springer.

Cloete, F. (2021). Science potential for development in Africa not yet fully realised. *South African Journal of Science, 117*(5/6), Art. #9390. https://doi.org/10.17159/sajs.2021/9390

Cooper, H., et al. (2023). Marketing communication education in developing countries: Post-pandemic insights from India and South Africa. *Journal of Marketing Communications, 30*, 1–23.

Dredge, D. et al. (2019). *Digitalisation in tourism: In-depth analysis of challenges and opportunities*. https://api.semanticscholar.org/CorpusID:211801433

Erol, I., Neuhofer, I. O., Dogru, T., Oztel, A., Searcy, C., & Yorulmaz, A. C. (2002). Improving sustainability in the tourism industry through blockchain technology: Challenges and opportunities. *Tourism Management, 93*, 104628.

Fu, S., Zheng, X., & Wong, I. (2022). The perils of hotel technology: The robot usage resistance model. *International Journal of Hospitality Management, 102*, 103174.

Garanti, Z., Zvaigzne, A., & Berjozkina, G. (2023). Introduction: Smart tourism: What developments and issues are important to the Baltic States? *Worldwide Hospitality and Tourism Themes*. https://doi.org/10.1108/WHATT-06-2023-0071

Goswami, M., & Daultani, Y. (2022). Make-in-India and industry 4.0: Technology readiness of select firms, barriers and socio-technical implications. *The TQM Journal, 34*(6), 1485–1505.

Govender, N., Laryea, S., & Watermeyer, R. (2022). Conceptual research methodology to investigate the relationship between fees and quality of professional service outputs. *Journal of Engineering, Design and Technology*. https://doi.org/10.1108/JEDT-11-2021-0629

Hughes, C., et al. (2022). Assessing the needs of the quantum industry. *IEEE Transactions on Education, 65*(4), 592–601.

Ivanov, S., & Webster, C. (2019). Economic fundamentals of the use of robots, artificial intelligence, and service automation in travel, tourism, and hospitality. In *Robots, artificial intelligence, and service automation in travel, tourism and hospitality* (pp. 39–55). Emerald Publishing Limited.

Jawabreh, O., Shniekat, N., Saleh, M., & Ali, B. (2022). The strategic deployment of information systems attributes and financial performance in the hospitality industry. *Information Sciences Letters, 55*(5), 110504.

Khatter, A. (2023). Challenges and solutions for environmental sustainability in the hospitality sector. *Sustainability, 15*(15), 11491.

Lamm, D. (2021). An exploration of environmental practices in the hospitality industry. Evidence from hotels in Addis Ababa, Ethiopia. *Journal of Environmental Management and Tourism (JEMT), 12*(2), 357–370.

Law, R., Buhalis, D., & Cobanoglu, C. (2014). Progress on information and communication technologies in hospitality and tourism. *International Journal of Contemporary Hospitality Management, 26*(5), 727–750.

Lee, P., Hunter, W., & Chung, N. (2020). Smart tourism city: Developments and transformations. *Sustainability, 12*(10), 3958.

Lu, C., et al. (2021). An energy-efficient smart city for sustainable green tourism industry. *Sustainable Energy Technologies and Assessments, 47*, 101494.

Madhav, A., & Tyagi, A. (2022). The world with future technologies (Post-COVID-19): Open issues, challenges, and the road ahead. In *Intelligent interactive multimedia systems for e-healthcare applications* (pp. 411–452).

Mapeshoane, T., & Pather, S. (2016). The adoption of e-commerce in the Lesotho tourism industry. *The Electronic Journal of Information Systems in Developing Countries, 75*(1), 1–24.

Mingotto, E., Montaguti, F., & Tamma, M. (2021). Challenges in re-designing operations and jobs to embody AI and robotics in services. Findings from a case in the hospitality industry. *Electronic Markets, 31*, 493–510.

Mnyakin, M. (2023). Big data in the hospitality industry: Prospects, obstacles, and strategies. *International Journal of Business Intelligence and Big Data Analytics, 6*(1), 12–22.

Moradi, E., Ehsani, M., Saffari, M., & Norouzi Seyed Hosseini, R. (2023). How can destination competitiveness play an essential role in small Island sports tourism development? Integrated ISM-MICMAC modelling of key factors. *Journal of Hospitality and Tourism Insights, 6*(3), 1222–1252.

Myovella, G., Karacuka, M., & Haucap, J. (2020). Digitalization and economic growth: A comparative analysis of Sub-Saharan Africa and OECD economies. *Telecommunications Policy, 44*(2), 101856.

Mzembe, A., Novakovic, Y., Melissen, F., & Kamanga, G. (2019). Institutional bricolage as an antecedent of social value creation in a developing country's tourism and hospitality industry. *Corporate Social Responsibility and Environmental Management, 26*(4), 997–1008.

Nabila, E. (2023). Digital transformation and AI in tourism: Trends, challenges, and successful experiences. *Economic Studies, 23*(1), 522–545.

Nam, K., et al. (2021). The adoption of artificial intelligence and robotics in the hotel industry: Prospects and challenges. *Electronic Markets, 31*, 553–574.

Ndhlovu, E., & Dube, K. (2023a). Challenges of radical technological transition in the restaurant industry within developing countries. *African Journal of Hospitality, Tourism and Leisure, 12*(1), 156–170.

Ndhlovu, E., & Dube, K. (2023b). Impact of COVID-19 on tourism in South Africa: Building resilience and prospects for recovery. In K. Dube, G. Nhamo, & M. P. Swart (Eds.), *COVID-19, tourist destinations and prospects for recovery volume one: A global perspective* (pp. 81–100). Springer Nature.

Nyagadza, B., Chuchu, T., & Chigora, F. (2022). Chapter 9: Technology application in tourism events: Case of Africa. In A. Hassan (Ed.), *Digital transformation and innovation in tourism events*. United Kingdom Tourism Society/Routledge.

Ogutu, H., et al. (2023). Theoretical nexus of knowledge management and tourism business enterprise competitiveness: An integrated overview. *Sustainability, 15*(3), 1948.

Otemoyolo, M. (2021). *Challenges and critical success of factors of business intelligence for small and medium Enterprises in the Republic of Congo*. Colorado Technical University.

Ozdemir, O., Dogru, T., Kizildag, M., & Erkmen, E. (2023). A critical reflection on digitalization for the hospitality and tourism industry: Value implications for stakeholders. *International Journal of Contemporary Hospitality Management*. https://doi.org/10.1108/IJCHM-04-2022-0535

Parsons, D., et al. (2023). The policy responses of tourism agencies to emerging digital skills constraints: A critical assessment of six countries. *International Journal of Tourism Research, 25*(1), 97–108.

Rodriguez Sanchez, I., Williams, A., & García Andreu, H. (2020). Customer resistance to tourism innovations: Entrepreneurs' understanding and management strategies. *Journal of Travel Research, 56*(3), 450–464.

Ruel, H., & Njoku, E. (2021). AI redefining the hospitality industry. *Journal of Tourism Futures, 7*(1), 53–66.

Tom Dieck, M., & Han, D. (2022). The role of immersive technology in customer experience management. *Journal of Marketing Theory and Practice, 30*(1), 108–119.

Van der Waldt, G. (2020). Constructing conceptual frameworks in social science research. *The Journal for Transdisciplinary Research in Southern Africa, 16*(1), a758. https://doi.org/10.4102/td.v16i1.758

Yallop, A., et al. (2023). The digital traveller: Implications for data ethics and data governance in tourism and hospitality. *Journal of Consumer Marketing, 40*(2), 155–170.

Yaqoob, I., Salah, K., Jayaraman, R., & Omar, M. (2023). Metaverse applications in smart cities: Enabling technologies, opportunities, challenges, and future directions. *Internet of Things, 23*, 100884.

Chapter 12
Leveraging Fintech Innovations for Post-COVID-19 Tourism Recovery in Asia: Lessons and Opportunities for Africa

David Mhlanga and Anupama Sethi

12.1 Introduction and Background

The tourism industry experienced a significant impact on a global scale because of the COVID-19 pandemic (Škare et al., 2021; Purba et al., 2021). Abruptly, airports that were formerly filled with activity were devoid of noise, renowned sites stood without the presence of tourists, and previously prosperous hotels and restaurants encountered the formidable task of enduring a period characterized by unpredictability. The tourism sector had significant challenges due to travel restrictions, border closures, and concerns about the spread of the contagion (Gössling et al., 2020; Seyfi et al., 2023). This has resulted in a precarious situation for numerous firms and employment opportunities within the industry. However, with the development and distribution of vaccines, coupled with the reopening of borders by governments, a sense of cautious hope started to surface. Individuals who engage in travel, fatigued by their restricted circumstances, and filled with a want for new experiences, began to envision exciting possibilities once more. The resurgence of a desire for travel, along with an increased emphasis on well-being and security, signified a pivotal moment for the tourism sector.

During this transformative period, the tourist sector has found a valuable friend in the form of Financial Technology (Fintech), commonly referred to as Fintech, to aid in its efforts to recover (Khayer & Alam 2022, Dwivedi et al. 2023). The

D. Mhlanga (✉)
College of Business and Economics, The University of Johannesburg, Johannesburg, South Africa

A. Sethi
Department of Economics, Monash Business School, Clayton, VIC, Australia
e-mail: anupama.sethi@monash.edu

E. Ndhlovu et al. (eds.), *Tourism and Hospitality for Sustainable Development*,
https://doi.org/10.1007/978-3-031-63069-9_12

integration of finance and technology, commonly referred to as fintech, has been progressively penetrating diverse sectors, resulting in the optimization of operational procedures, bolstering of security measures, and facilitation of financial service accessibility. Within the realm of tourism, the role of Fintech has assumed a novel function, namely that of a catalyst for the process of recovery. It became apparent that the sector needs novel solutions to effectively adjust to the evolving terrain. Fintech has presented a range of tools that have the potential to revolutionize travel experiences in a post-pandemic era, including touchless payment methods, digital identity verification, and blockchain-based travel records. The purpose of this chapter is to investigate how Fintech is linked to the tourism industry, with a specific emphasis on its potential to bring about significant changes for Asia's post-COVID recovery. Additionally, we will explore the lessons that Africa can gain from this process. This study aims to delve into the possibilities offered by Fintech solutions in enhancing efficiency and security within the travel sector using a systematic literature review. By analysing how these solutions impact various aspects of travel, such as booking and managing expenses after the trip, our research aims to uncover how Fintech can contribute to the industry's recovery following the disruptions caused by the pandemic.

12.2 Methods and Data

A systematic literature review is a meticulous and all-encompassing examination of academic papers, reports, and other pertinent materials about a certain subject, query, or field of study. The primary objective is to ascertain, assess, and amalgamate all the existing research findings about a certain query or field of interest (Mhlanga, 2023a; Sandelowski & Barroso, 2006). This methodological approach distinguishes itself from conventional literature reviews by utilizing a more systematic and clear-cut methodology for doing research, choosing relevant studies, and evaluating their quality. Below is a concise overview of the essential stages in performing a systematic literature review: Defining a Research Question, creating a Protocol, carrying out an Exhaustive Search, and Screening and Choosing Studies. The process involves extracting data, evaluating its quality, synthesizing the findings, and finally reporting and disseminating the results. The produced report adheres to the PRISMA guidelines or other applicable standards for systematic reviews and meta-analyses. The report must encompass the background, methodologies, findings, discourse, and conclusions of the review. Another phase involves updating the review.

12.3 Review of Important Literature

12.3.1 The Meaning of Fintech

Fintech, a term derived from combining "Financial" and "Technology", encompasses the innovative use of technology to upgrade and automate the delivery and use of financial services and processes (Mhlanga, 2020, 2022a, Puschmann, 2017). The convergence of finance and technology has revolutionized the financial landscape by reshaping payments, lending, investment, insurance, and other financial products and services (Feyen et al., 2023; Goldstein et al., 2019; Morgan, 2022). Companies within the Fintech sector employ advanced software, applications, and digital platforms, focusing on enhancing efficiency, accessibility, and affordability across a myriad of financial activities, ensuring streamlined transactions and services for both businesses and consumers (Mhlanga, 2022b, 2023b). It leverages software, applications, and digital platforms to provide digital financial services through devices like smartphones, and is increasingly acknowledged as a powerful means to foster financial inclusion (Mhlanga 2021, 2022).

Many Fintech firms have emerged for instance, digital payment providers (e.g. PayPal), and peer-to-peer lending platforms (e.g. Afluenta, Funding Circle, Investree) to facilitate financial transactions. Most Fintech have incorporated open banking standards in their products and services, which allows seamless provision of financial services using aggregated and authenticated customer data ("Five key Fintech trends to watch in 2022 according to Buckzy Payments", 2021). Fintech has the potential to enhance transparency and reduce information asymmetries (Cong & He, 2019; Feyen et al., 2023). Technologies such as artificial intelligence, machine learning, blockchain, cloud computing, the Internet of Things, and big data have enabled and contributed towards the advancement of financial segments, for better transaction and security (Awotunde et al., 2021). Furthermore, technology is playing a crucial role in lowering transaction costs by overcoming geographical access barriers, enhancing the speed, security, and transparency of financial transactions, and allowing for more personalized financial services that cater to consumers' needs (Demirgüç-Kunt et al., 2018; Feyen et al., 2023). According to Demirgüç-Kunt et al., (2022), there is a significant global expansion of financial inclusion. Its statistics show the percentage of adults engaging in digital payments in developing economies surged from 35% in 2014 to 57% in 2021.

Examples of Fintech Innovations around the Globe, M-Pesa (Kenya), a globally recognized mobile money platform, that facilitates payments, money transfers, and access to financial services via mobile phones. It has significantly impacted financial inclusion not only in Kenya but also in other regions it serves. Flutterwave platform (Nigeria) provides comprehensive payment solutions for businesses, enabling them to process payments, manage transactions, and access other financial services both within Africa and on an international scale. Paystack aids businesses in Nigeria and across Africa by providing a platform that allows them to accept payments online and offline, thereby smoothening their financial operations and customer transactions. Cellulant (Kenya), as a pan-African digital payments platform, connects businesses and governments with consumers through diverse digital channels, offering services like digital wallets and digital payments to streamline financial activities. Yoco (South Africa) provides point-of-sale (POS) solutions to small businesses, empowering them to accept card payments using mobile devices like smartphones or tablets, thereby enhancing financial access for small-scale merchants. Other International Digital Payment Platforms, include PayPal enabling secure online payments and money transfers for individuals and businesses, PayPal has positioned itself as a global leader in the Fintech sphere. Square offers adaptable point-of-sale systems and mobile card readers, focusing on providing small businesses with accessible and user-friendly financial management solutions. Venmo is popular for its ease in enabling peer-to-peer money transfers and mobile payments, Venmo provides a social way for friends and family to send and receive money seamlessly. Fintech has unarguably revolutionized the financial landscape globally, driving advancements that enable easier transactions, broader financial access, and innovative solutions to traditional financial issues. The companies exemplify the diverse applications and widespread impact of Fintech across different regions and markets.

12.3.2 Fintech and Tourism

The tourism sector has witnessed notable advancements from the integration of fintech, which has introduced inventive measures to improve the travel experience for both tourists and stakeholders within the industry (Buhalis et al., 2019; Kaurav et al., 2020). Fintech is employed in several capacities within the tourism industry. The utilization of online booking and reservation systems is of great significance in the realm of fintech. These systems enable travellers to securely make bookings for various travel-related services such as flights, lodgings, tours, and activities by leveraging robust online payment systems. Prominent online travel agencies (OTAs), such as Expedia, Booking.com, and Airbnb, employ fintech to enable seamless payment transactions and enhance the efficiency of the booking procedure. In the wake of the COVID-19 pandemic, the significance of contactless payment systems has been underscored, mostly due to their emphasis on hygiene and safety in the current era. Fintech facilitates the utilization of mobile wallets such as Apple Pay and Google Pay, contactless cards, and QR code-based payment systems by travellers, hence diminishing the reliance on traditional methods of cash or card transactions while travelling. Certain travel companies and platforms have recently begun to acknowledge cryptocurrencies, such as Bitcoin, as a viable means of transaction (Baralla et al., 2021; Teichmann & Falker, 2021). This feature offers a supplementary method of payment for technologically proficient individuals who are travelling, thereby aiding in the smooth execution of transactions that occur across national borders.

The implementation of digital identity verification and biometric authentication in the field of fintech has facilitated secure and streamlined check-in procedures at various establishments including airports, hotels, and automobile rental companies. The implementation of this measure results in an improvement in security measures and expedites the process of verification, hence lowering the duration of waiting periods for individuals travelling. The integration of fintech has significantly enhanced the efficiency of travel insurance policy procurement and claims processing. Travellers can acquire insurance through online platforms, and the utilization of automated claims processing algorithms can facilitate the prompt disbursement of compensation in instances of travel-related predicaments such as delays, cancellations, or other similar occurrences. Fintech platforms provide currency exchange services with competitive exchange rates and minimal transaction fees. Travellers have the option to utilise mobile applications and online platforms for currency conversion before or during their journeys, hence obviating the necessity of physical exchange counters.

Expense management is facilitated by fintech tools and applications, which aid travellers in monitoring and controlling their expenditures throughout their journeys. These solutions offer real-time updates regarding spending, provide currency conversion functionality, and categorize expenses to facilitate budget management. The Impact of Loyalty Programmes and Rewards on Consumer Behaviour Numerous entities within the travel industry employ fintech solutions to effectively administer loyalty programmes and incentives. Travellers could accrue and utilize points, miles, or discounts on airfare, lodging, and various other expenses associated with travel employing mobile applications and internet platforms. The utilization of blockchain technology in the context of travel records. The potential application of blockchain technology is now being investigated in the development of safe and unalterable travel records, encompassing digital passports and vaccination certificates. These records have the potential to enhance the efficiency and security of international travel. The integration of personalized financial services within the realm of Fintech has facilitated the provision of customized financial advice and services that are specifically designed to cater to the unique requirements of individuals who engage in travel activities.

Mobile banking applications offer a range of functionalities, such as expense monitoring, budget management, and instantaneous transaction notifications, which serve to assist individuals in effectively overseeing their financial matters during their travels overseas (Mhlanga, 2023a, b). Peer-to-peer lending has emerged as a viable alternative for those seeking financial assistance to fund their travel endeavours. Certain fintech platforms provide peer-to-peer lending services, facilitating the provision of loans to travellers to cover their travel-related expenditures. These loans possess a higher degree of flexibility and accessibility compared to conventional bank loans. The utilization of digital wallets to manage travel expenses. Digital wallets such as PayPal, as well as mobile payment applications like Apple Pay and Google Pay offer travellers the convenience of expeditious and secure payment options for a wide range of travel-related expenditures, encompassing meals and transportation. Fintech advancements persistently reshape the tourism sector, resulting in increased convenience, security, and accessibility for travellers, alongside operational streamlining and improved customer experiences for firms operating within the industry.

12.3.3 The Significance of Tourism in Both Asian and African Economies

Tourism plays a significant role in many economies worldwide and its importance is particularly noticeable in the diverse regions of Asia and Africa. Tourism is a powerful economic engine, driving growth and job creation across the globe. In the vibrant and culturally rich continents of Asia and Africa, its impact is especially pronounced. In both regions, tourism serves not just as an economic pillar, but as a bridge fostering global understanding and appreciation.

Asia, with its blend of ancient traditions, varied landscapes, and bustling modern cities, attracts millions, boosting local economies, promoting cultural exchange, and enhancing infrastructure. Figure 12.1 illustrates the economic contribution of tourism in Asian economies.

Fig. 12.1 Asian economies and economic contribution from tourism

Direct Revenue and Employment

Tourism is a linchpin in the economic machinery of countries like Thailand, Malaysia, Indonesia, and the Maldives, funnelling substantial foreign exchange into their coffers (Adeleye et al., 2022a, b). These nations have seen their natural beauty and cultural attractions transform into bustling tourist hubs, creating myriad employment opportunities across sectors. Bali, with its enchanting landscapes and rich culture, stands as Indonesia's crown jewel, with its economy deeply intertwined with tourism. Similarly, the Maldives, with its constellation of luxury resorts scattered across its turquoise atolls, not only draws high-end travellers but also provides a livelihood for a significant segment of its population, showcasing the intricate nexus between tourism, revenue, and employment in these nations.

Infrastructure Development and the Promotion of Ancillary Services

The surge in tourism often acts as a catalyst for infrastructure enhancement, ushering in advanced and more efficient facilities to accommodate the influx. For example, the rapid growth of tourism in Dubai led to the expansion of its already impressive airport, further solidifying its position as a global travel hub. Similarly, in Vietnam, the allure of places like Da Nang and Hoi An has fostered significant upgrades in road networks and the construction of modern airports. These infrastructural boosts, while catering to tourists, also provide local businesses with improved logistics and connectivity, while offering residents enhanced amenities and transportation options, thereby elevating the overall quality of life in the region.

Tourism doesn't just boost hotel and travel industries. It promotes growth in ancillary services like food, entertainment, retail, and handicrafts. Tourism's economic ripples extend far beyond the obvious realms of hotel bookings and transport services, creating a thriving ecosystem of ancillary businesses that cater to the diverse needs and tastes of visitors. The bustling markets of Bangkok, for instance, don't merely offer souvenirs but a tantalizing array of street food, exemplifying how culinary experiences become an integral part of a traveller's journey. Similarly, tourists flocking to the serene backwaters of Kerala in India don't just seek boat rides; they often indulge in local Ayurvedic spa treatments, demonstrating the symbiotic relationship between tourism and local health and wellness sectors. Moreover, the entertainment sectors of many destinations have seen rejuvenation owing to tourism. For example, the traditional Balinese dances in Indonesia or the Maasai performances in Kenya have found enthusiastic audiences in tourists, ensuring these art forms remain vibrant and lucrative. Local handicrafts, too, get a significant boost. Tourists visiting Marrakech, Morocco, often leave with intricately designed rugs or pottery, directly supporting artisans and ensuring the continuation of these crafts for future generations. This interplay between tourism and ancillary services is a testament to the holistic impact of tourism on local economies.

Cultural Exchange and Diversification

Tourism in Asia provides a platform for cultural exchange, promoting mutual respect and understanding. Countries like Japan, China, and India, with rich histories, benefit from this interaction. Tourism acts as a bridge, linking diverse cultures and fostering a tapestry of shared experiences and mutual appreciation. In Asia, where ancient traditions and histories abound, this cultural exchange is vividly evident. For instance, when visitors partake in Japan's traditional tea ceremonies, they immerse themselves in a ritual that goes beyond mere tasting, delving deep into the nation's philosophical and aesthetic ideals. Likewise, tourists exploring China's Great Wall not only marvel at its architectural prowess but also engage with millennia of history and strategic ingenuity. Similarly, India's Diwali festival, known as the Festival of Lights, draws global visitors who participate in its vibrant celebrations, understanding the profound spiritual and cultural significance attached to it. Through these interactions, both visitors and locals gain insights into each other's cultures, breaking down barriers and fostering a sense of global community.

Some Asian economies which have been traditionally dependent on agriculture or specific industries are looking at tourism as a means to diversify their economic bases. For example, Vietnam has seen a rise in tourism-related activities over the years. Diversification is pivotal for economies to ensure resilience against market fluctuations, and tourism offers a robust avenue for this purpose in Asia. Countries historically reliant on specific sectors have recognized the potential of tourism to broaden their economic horizons. Vietnam, for instance, once primarily an agrarian economy, has tapped into its rich cultural tapestry and breathtaking landscapes, from the bustling streets of Ho Chi Minh City to the serene waters of Halong Bay, attracting a multitude of tourists. This influx has spurred a slew of related industries, from hospitality to local crafts. Similarly, Cambodia, once primarily known for its rice production, has seen its ancient temples like Angkor Wat transform into global tourist magnets, thereby infusing fresh streams of revenue and offering new employment opportunities, underlining the transformative power of tourism-led diversification. Africa's unique tapestry of wildlife, landscapes, and cultural heritage has made it a sought-after destination. Tourism revenue here aids conservation efforts, supports local communities, and often acts as a bulwark against economic challenges (Büscher & Fletcher 2019; Higgins-Desbiolles 2022). Figure 12.2 illustrates the economic contribution of tourism in African economies.

Direct Revenue and Employment

Tourism in countries such as Kenya, Tanzania, South Africa, and Egypt not only generates substantial direct revenue but also significantly propels employment across various sectors. The majestic safari tours in Kenya and Tanzania, for instance, offer an unparalleled wildlife experience, drawing thousands annually, thereby fuelling local employment from safari guides to campsite staff. Egypt, with its awe-inspiring pyramids and ancient temples, sees a consistent flow of history enthusiasts,

Fig. 12.2 African economies and tourism economic contribution

leading to job creation in sectors ranging from museum management to local guiding. Meanwhile, South Africa, boasting a diverse topography from Table Mountain to its sprawling vineyards, attracts a gamut of tourists, subsequently promoting job growth in hospitality, wine production, and adventure sports. Collectively, these nations exemplify how tourism, by luring global audiences, can be a cornerstone for both revenue generation and job creation. Further delving into the intricate dance of tourism and its impacts on economies, we witness a multifaceted interplay of influences.

In Kenya, apart from the famed Masai Mara National Reserve, coastal regions like Mombasa cater to beach enthusiasts, indirectly boosting local businesses, from seafood restaurants to handicraft stalls. Tanzania's Zanzibar, with its historic Stone Town and spice tours, is an added feather in the nation's cap, providing livelihoods to spice farmers and artisans alike. South Africa's Garden Route, stretching between Mossel Bay and Storms River, is another testament to the country's vast tourist offerings. This scenic drive promotes numerous B&B establishments, adventure sports organizers, and boutique eateries, all contributing to the local job market. Moreover, its vibrant cities like Johannesburg and Cape Town have seen an upsurge in urban tourism, supporting everything from boutique artisans in local markets to operators of township tours. Egypt's tourism isn't confined just to its ancient relics. The Red Sea destinations like Sharm el-Sheikh have become a hub for diving enthusiasts, leading to a proliferation of dive shops, resorts, and related services, adding another dimension to Egypt's tourism-driven employment. In essence, the magnetic allure of these countries doesn't just stem from their high-profile attractions. The intricate web of services, experiences, and interactions that tourists engage with often behind the scenes significantly amplifies the direct revenue and employment opportunities in these African gems.

Boost to Local Communities

Community-based tourism (CBT) in African nations is a transformative approach that embeds tourism within the very fabric of local communities, ensuring they are not mere spectators but active participants in the sector. This model prioritizes the welfare, culture, and aspirations of these communities, directing tourism revenue to the hands of those who need it most. For instance, in Namibia, the conservancy program has seen local communities take the reins of wildlife management and tourism. These conservancies, such as the Torra Conservancy, have not only led to significant wildlife recovery but have also provided the local Himba and Herero populations with sustainable incomes from tourism, be it through guiding, accommodation, or cultural experiences. The ripple effect of this is profound, as it leads to improved infrastructure, education, and health services for the community. Kenya provides another exemplary model with its Maasai-owned and managed conservancies in the Mara region. These conservancies, such as Naboisho or Olare Motorogi, ensure that Maasai landowners earn directly from tourism activities, while also being actively involved in conservation efforts. This not only empowers the community economically but also preserves their age-old traditions, as tourists get an authentic insight into Maasai culture, from village visits to interactive storytelling sessions. Furthermore, the revenue generated often goes back into projects of local significance, like schools or water supply systems, painting a holistic picture of how community-based tourism can be a catalyst for grassroots development.

Conservation Financing and Promotion of African Heritage

In the intricate dance of conservation and commerce in Africa, tourism often takes centre stage as a sustainable funding tool. The vast expanses of national parks, wildlife reserves, and conservancies across the continent rely heavily on the influx of tourists, with entrance fees, accommodation charges, and guided tours becoming instrumental in funding conservation endeavours. For instance, Rwanda's Volcanoes National Park, home to the critically endangered mountain gorillas, uses its revenue from gorilla trekking permits priced at several hundred dollars each to support conservation efforts. These funds directly contribute to anti-poaching measures, habitat restoration, and community-based projects that aim to reduce human-wildlife conflict, ensuring the survival of these majestic creatures. Similarly, Botswana, which houses some of Africa's most pristine wilderness areas, such as the Okavango Delta and Chobe National Park, incorporates a high-value, low-impact tourism model. This approach focuses on attracting fewer tourists willing to pay a premium for an authentic and exclusive experience. The resulting revenue then funnels directly into the maintenance of these protected areas, supporting initiatives like anti-poaching units, wildlife monitoring, and research programs. Furthermore, these funds also facilitate community engagement in conservation, bridging the gap between local communities and the wildlife they coexist with, ensuring a future where both can thrive in harmony.

Tourism helps promote African art, culture, music, and traditions on a global stage. Tourism in Africa serves as a vibrant showcase for the continent's rich tapestry of art, culture, music, and traditions, presenting them to a global audience and ensuring their preservation and proliferation. The bustling markets of Marrakech, Morocco, for instance, introduce tourists to traditional Berber crafts, from intricate rugs to ornate lanterns. Similarly, the rhythms and melodies of Africa are echoed in the Senegalese drum circles or the spirited dance performances in Ghana, offering a captivating musical journey to visitors. The ancient rock-hewn churches of Lalibela, Ethiopia, and the vibrant Maasai dances of Kenya, on the other hand, provide a deep dive into the continent's religious and tribal rituals. Through these interactions, tourism not only sustains the livelihoods of countless artisans, musicians, and performers but also amplifies the diverse voices of African heritage, fostering appreciation and understanding on an international scale.

Diversification

Just as in Asia, tourism provides a chance for African economies to diversify beyond traditional industries, whether that's agriculture or mineral extraction. In the vibrant continent of Africa, tourism emerges as a potent catalyst for diversification, enabling countries to augment their economic portfolios beyond traditional mainstays. Many African nations, historically reliant on sectors like agriculture or mineral extraction, now view tourism as a golden opportunity to diversify their income sources, thereby mitigating risks associated with over-reliance on a single sector. Take Zambia, for instance. While copper extraction remains a dominant industry, the country has increasingly turned to its natural wonders, like the majestic Victoria Falls, to boost its tourism sector. With activities ranging from waterfall tours to adrenaline-pumping bungee jumps, Zambia now sees a steady stream of tourists, leading to the sprouting of hotels, restaurants, and ancillary services, diversifying its once mineral-centric economy. Similarly, Senegal, traditionally an agrarian economy with major exports like peanuts and fish, has recognized the potential of its sun-kissed beaches, rich history, and vibrant culture. Destinations like Goree Island, a UNESCO World Heritage site, and the Pink Lake have become significant tourist draws. As a result, the tourism sector has not only provided an alternative source of revenue but has also spawned a myriad of related industries, from handicraft vendors to local tour operators. By intertwining their unique attractions with global interests, these nations exemplify the transformative role of tourism in reshaping and diversifying economic landscapes.

12.3.4 Challenges of Over-tourism

While tourism offers immense economic potential, it also brings forth significant challenges, particularly when not managed sustainably. Over-tourism, a phenomenon where an overwhelming number of tourists inundate a destination, can lead to environmental degradation and strain local resources. Boracay, a pristine island in the Philippines, is a case in point. Its white-sand beaches and turquoise waters attracted such a surge of visitors that it led to environmental concerns, including waste management issues and threatened marine ecosystems. As a result, the island had to be temporarily closed for rehabilitation in 2018 (Benaraba et al. 2022; Khan & Hashim 2020). Similarly, Venice in Italy, with its historic canals and architecture, faces threats from rising waters and the sheer volume of tourists, leading to concerns about the city's long-term viability as a travel destination. These instances underscore the importance of balancing tourism growth with sustainable practices to preserve the very attractions that draw visitors in the first place.

Challenges of Tourism in Africa

Tourism in Africa often faces challenges like political instability, health concerns (e.g. the Ebola outbreak), and infrastructural issues. The potential of Africa's tourism sector, while vast, is frequently tempered by an array of challenges that can impede its consistent growth and development. Political instability, often characterized by conflicts, civil unrest, or frequent regime changes, can deter potential tourists and investors. For example, the magnificent ruins of Great Zimbabwe, a testament to an ancient African kingdom, or Mali's historic city of Timbuktu, with its centuries-old manuscripts, have often seen reduced footfall during periods of political unrest in their respective countries. Such instability not only casts a shadow over immediate tourist arrivals but also hampers long-term investment in the sector, affecting everything from hotel construction to tour operations. Health concerns, too, pose significant challenges. The Ebola outbreak that began in 2014 in West Africa, primarily affecting countries like Sierra Leone, Guinea, and Liberia, had a ripple effect across the continent's tourism industry. Even nations distant from the epicentre of the outbreak experienced dwindling numbers of visitors due to generalized fears and misconceptions. Meanwhile, infrastructural issues, ranging from inadequate road networks to limited air connectivity, can also hamper the ease of travel. A tourist keen on experiencing both the Sahara's vastness in Algeria and the lush landscapes of Uganda's Bwindi Impenetrable Forest might find the journey logistically challenging. Such infrastructural gaps necessitate significant investments to unlock the full potential of Africa's diverse attractions. While the continent offers an unmatched wealth of experiences, addressing these challenges is crucial for the sustained growth of its tourism industry.

12.3.5 *Benefits of Integrating Fintech in Africa's Tourism Sector*

The main potential benefits of Fintech in Africa's tourism sector are increased accessibility and enhanced customer experience, cost-efficiency and financial inclusion, and real-time data analysis and security (Fig. 12.3).

Increased Accessibility and Enhanced Customer Experience

The integration of financial technology (fintech) in the tourism industry has facilitated the democratization of travel, allowing individuals with varying financial resources to efficiently manage their expenses. The advent of platforms such as mobile money has brought about a significant transformation in payment systems, hence reducing the reliance on credit cards as the exclusive means of conducting financial transactions. M-Pesa, a mobile payment system in Kenya, serves as a notable example of this societal shift. It enables seamless access and payment for a wide range of services, including accommodations and recreational activities, for both domestic and international users. This method circumvents the need for conventional cash or card-based transactions. Not only does this streamline processes, but

Fig. 12.3 Benefits of integrating Fintech in Africa's tourism sector

it also expands the possibilities for prospective travellers. The integration of AI and machine learning into the tourism sector has profoundly elevated the customer experience. These advanced technologies allow digital platforms to discern and anticipate user preferences, delivering highly personalized travel recommendations. Gone are the days of cumbersome physical bookings and the drudgery of waiting in lines. Platforms such as Jumia Travel in Africa exemplify this evolution, offering AI-powered hotel suggestions rooted in a user's search and reservation history. Such innovations ensure a more intuitive and user-centric approach to travel planning, streamlining processes and aligning offers more closely with individual desires.

Cost-Efficiency and Financial Inclusion

With digital transactions, businesses can cut down on costs related to cash management, such as safes, armoured vehicles, and teller wages. These savings can then be redirected to improve services or reduce prices. A hotel that adopts digital payment solutions might save on bank transaction fees and related costs. These savings can potentially be passed on to consumers in the form of discounted rates or special offers. Many Africans remain unbanked, yet a vast majority now have mobile phones. Mobile financial solutions can bring banking and financial services to this previously unreachable population. In places like rural Ghana, where traditional banking is scarce, services like MTN Mobile Money offer local vendors and tourists alike the opportunity to transact digitally, ensuring smooth commerce in tourism hotspots.

Real-Time Data Analysis and Security

Digital transactions provide an influx of data that can be analysed. This real-time data can inform businesses about peak visiting times, and popular services, and even forecast future trends. A safari company in Tanzania might notice a spike in bookings during a particular season through its digital booking system. This information can be used to hire seasonal staff, offer special packages, or prepare for increased visitors during those periods. Digital payments reduce the risks associated with cash theft or fraud. Blockchain, a fintech solution, provides immutable ledgers, ensuring transaction transparency and security. A luxury resort in South Africa accepting digital payments can assure its customers that their financial data is secure, as transactions are encrypted and often require two-factor authentication. Additionally, platforms that utilize blockchain, like some travel booking sites, can offer guests transparency in pricing and services, ensuring they receive what they've paid for. In essence, the integration of Fintech in Africa's tourism sector not only offers convenience and efficiency but also democratizes access, ensuring that the continent's vibrant cultures, landscapes, and experiences are more accessible to all.

12.3.6 Drawing Parallels Between Asian and African Countries in Terms of Tourism Challenges and Opportunities

Infrastructure Development in Asia and Africa

Both Asia and Africa, two vast continents with diverse landscapes, cultures, and histories, have grappled with challenges in the realm of infrastructure development to support and boost tourism. While both continents have unparalleled natural beauty and cultural attractions, the disparity in infrastructure investment and its resultant outcomes is evident. Asia, particularly nations such as Thailand and Malaysia, has recognized the vast potential of tourism as an economic driver. These countries have channelled significant resources into developing world-class transportation networks, accommodation facilities, and other amenities crucial for catering to international tourists. As a result of these strategic investments, places like Bangkok, Kuala Lumpur, Phuket, and Penang have witnessed a phenomenal surge in visitors, making tourism one of their primary revenue sources. On the other hand, many African countries, despite their equally enticing attractions, are yet to fully realize and capitalize on their tourism potential due to limited infrastructure development. These disparities underscore the importance of consistent investment and planning in infrastructure to truly harness the economic benefits of tourism.

Cultural Preservation in Asia and Africa

Asia and Africa, each boasting millennia of history, are treasure troves of rich cultural heritages, holding stories of ancient civilizations, traditions, and practices. In Asia, locales like Bali in Indonesia stand as examples of how regions can merge the advancements of modernity without overshadowing their intrinsic cultural values. Bali's delicate balance is evident in its bustling tourist hotspots which seamlessly coexist alongside ancient temples and time-honoured rituals, ensuring that visitors experience both the contemporary vibrancy and the profound depths of Balinese culture. Meanwhile, Africa, a continent renowned for its kaleidoscope of languages, ethnicities, and traditions, has a vast reservoir of cultural experiences waiting to be showcased through tourism. From the Maasai dances in the East to the rich tapestry of storytelling in the West, the potential is immense. However, this potential remains largely untapped, not due to a lack of cultural richness, but often because of insufficient avenues and platforms to present it. For both continents, the challenge lies in navigating the complex interplay between cultural preservation and the demands of modern tourism, ensuring that in the quest for economic growth, the soul of their civilizations is not lost.

Economic Disparities in Tourism and the Role of Fintech

Asia and Africa, despite their expansive landscapes and vast potentials, grapple with pronounced economic disparities, especially in the context of tourism. Certain destinations, endowed either by nature or history, become magnets for tourists, subsequently drawing substantial investment and infrastructure development. These areas, such as Bali in Asia or Cape Town in Africa, often experience rapid economic growth, enhancing living standards and creating numerous opportunities for the local populace. Conversely, adjacent areas, sometimes just as rich in cultural or natural beauty, might languish in obscurity, receiving negligible attention and development funds. This uneven distribution leads to stark contrasts in economic well-being even within short geographic distances. Fintech, or financial technology, presents an innovative avenue to address these disparities. By harnessing digital tools, platforms, and services, Fintech can facilitate smoother financial transactions, enabling businesses in less-touristic regions to access loans or crowdfunding, thus spurring localized development. Additionally, it can aid in the direct and transparent distribution of tourism revenues, ensuring that even the most remote service providers get their fair share. Through leveraging Fintech solutions, there is potential to create a more inclusive and equitable tourism ecosystem, mitigating the glaring economic contrasts that currently exist between popular and lesser-known destinations.

12.3.7 Empirical Literature Fintech in Tourism Recovery Post-COVID-19

Empirical literature refers to scholarly works that are based on observation, experimentation, or data analysis. These works are grounded in empirical evidence The Role of Financial Technology in Facilitating the Recovery of the Tourism Industry. The post-COVID-19 landscape is experiencing gradual expansion but with certain constraints and occasional indirectness. The phenomenon of post-COVID-19 is currently seeing growth but with certain limitations and occasional indirectness. In their study, Rabbani et al. (2020) constructed a model that highlights COVID-19 as a potential avenue for enhanced integrated and transformative growth, characterized by a heightened emphasis on standardization and the industry's concerted efforts towards addressing social issues, with the strategic implementation of financial technology. Rabbani et al. (2020) believe that the novel coronavirus (COVID-19) is a phenomenon that will have enduring consequences, which will be experienced in the forthcoming years. The Islamic banking business has been significantly impacted by the economic repercussions of the epidemic, as evidenced by the subdued growth projections put forth by numerous authorities. According to Rabbani et al. (2020), the current pandemic presents a favourable circumstance for the Islamic banking sector to expand and demonstrate its value once more, following the global financial crisis of 2008, thereby establishing itself as a significant competitor to the conventional financial system. The research conducted by Rabbani et al. (2020) holds significant implications for the Islamic finance and banking sector, as it offers a comprehensive framework that can guide future scholars and professionals in comprehending and implementing Islamic financial practices in the aftermath of the COVID-19 pandemic.

In their study, Lontchi et al. (2023) investigated the potential mediating role of financial literacy in the relationship between fintech and the performance of small and medium enterprises (SMEs) in Cameroon, specifically focusing on the recovery period following the COVID-19 pandemic. The research conducted by Lontchi et al. (2023) demonstrates a positive and statistically significant association between fintech and financial literacy. Furthermore, the study reveals that fintech exerts a positive and statistically significant impact on the performance of small and medium-sized enterprises (SMEs). Furthermore, the study conducted by Lontchi et al. (2023) revealed a strong and favourable mediation effect of financial literacy on the link between fintech services and the performance of small and medium-sized enterprises (SMEs). According to the findings of Lontchi et al. (2023), financial literacy plays a crucial role in mediating the impact of fintech services on the operational and financial performance of small and medium-sized enterprises (SMEs) in Cameroon. The study conducted by Lontchi et al. (2023) contributes to the existing body of knowledge by offering a more comprehensive insight into the potential for small and medium-sized enterprises (SMEs) to enhance their performance through the use of financial technology (fintech) solutions, specifically utilizing enhancing their financial literacy.

In their study, Vena-Oya et al. (2022) utilized a fuzzy approach to create a methodology aimed at generating forecasting scenarios that are both consistent and dependable. The objective of this methodology is to support decision-making processes for managers and policymakers, particularly in situations characterized by complexity and uncertainty. According to Vena-Oya et al. (2022), the utilization of a semi-quantitative model can assist researchers in predicting the potential consequences of significant occurrences within ambiguous or unclear settings through the development of versatile and adaptive scenarios. According to Park et al. (2022), the tourism and hospitality industries are undergoing significant transformations as a result of the COVID-19 pandemic. However, there is a scarcity of research examining the alterations in travel behaviour that have occurred in reaction to this global health crisis. Park et al. (2022) employ network science analytics to examine the alterations in travel mobility patterns of domestic tourists who travelled to Jeju Island, Korea, throughout the period from June 2019 to December 2020, utilizing tourism big data. According to the research conducted by Park et al. (2022), there were notable decreases in the frequency of visits to a particular location during the pandemic. Additionally, the study found that there were constraints on the extent of geographical interconnectedness and variety in travel patterns. According to the study conducted by Park et al. (2022), there is evidence to suggest that there is a higher intensity of travel mobility to outdoor and coastal locations, as well as shorter journey lengths, observed during the COVID-19 pandemic.

In their study, Suyunchaliyeva et al. (2021) examined the potential of mobile payment services to facilitate the recovery of the "experience goods" sector, specifically in the context of tourism, following the COVID-19 epidemic. The researchers employed Bandura's self-efficacy or social cognition theory as a theoretical framework for their investigation. The study conducted by Suyunchaliyeva et al. (2021) examined the determinants that impact individuals' inclination to persist in utilizing mobile payment services, as well as their inclination to endorse these services to others. According to the findings of Suyunchaliyeva et al. (2021), there is a relationship between personal innovativeness and perceived trust, and how they impact consumers' intention to continue using mobile payment services. Additionally, the study suggests that perceived trust, personal innovativeness, and outcome expectancy are factors that influence consumers' intention to recommend the use of mobile payment services to others. According to the findings of Leung et al. (2023), virtual reality (VR) technology has been identified as a viable instrument for tourism marketers to sustain the appeal of their locations and facilitate recovery in the aftermath of the COVID-19 pandemic. Nevertheless, the efficacy of virtual reality (VR) technology in stimulating the inclination of prospective tourists to visit destinations during periods of lockdown has yet to be determined, as posited by Leung et al. (2023).

Leung et al. (2023) employed an integrated model that draws upon the experience economy framework and mood management theory to elucidate the impact of tourists' virtual reality (VR) experiences on their mood management processes and subsequent behaviours. Leung et al. (2023) conducted a study that investigated the impact of perceived travel risk on the association between mood management procedures and subsequent decision-making. The findings presented by Leung et al. (2023) indicate that the four dimensions of virtual reality (VR) have varying effects on mood management processes. Additionally, the perceived danger associated with virtual travel has a moderating effect on how mood management processes influence visit intention and VR stickiness. Leung et al. (2023) offer valuable insights for tourism marketers to effectively navigate the present tourist landscape and formulate recovery plans.

The objective of the study conducted by Amin and Taghizadeh-Hesary (2023) was to examine the interconnections between tourism, sustainability, and innovation within the context of the Bangladeshi economy. According to Amin and Taghizadeh-Hesary (2023), an analysis was conducted using yearly data from 1980 to 2019. The researchers employed a standard and widely accepted econometric method for time series analysis. The findings of their study indicate a substantial association between innovation, sustainability, and the tourism industry in Bangladesh. Moreover, the influence of the COVID-19 pandemic on the tourism business was investigated by Amin and Taghizadeh-Hesary (2023) through the utilization of a computable General Equilibrium (CGE) modelling approach. It has been observed that the tourism sector in Bangladesh experiences significant impacts. According to Amin and Taghizadeh-Hesary (2023), the implementation of innovation efficiency and government stimulus packages emerges as a viable recovery method for addressing the negative consequences of the epidemic. Amin and

Taghizadeh-Hesary (2023) offer significant policy recommendations for the sustainable growth of the tourism sector, drawing upon a comprehensive analysis. Having a closer look at all the studies reviewed none of them looked at the relationship of FinTech and Tourism recovery.

The Rise of Fintech in Asia Post-COVID-19

The COVID-19 pandemic has expedited the adoption and spurred further innovation in Fintech (Feyen et al., 2023; Morgan, 2022). The global pandemic ushered disruption for businesses across the globe, with the tourism industry highly affected. In response to these new challenges, the tourism and hospitality industry has harnessed the advancements in the Fintech sector (Toumi et al., 2023). The pandemic accelerated the shift towards contactless payment methods. A few challenges during the pandemic travelling were due to ever-changing rules during the pandemic. Thus, the customer's preferences for flexible, contactless financial transactions increased during the pandemic. Fintech firms have introduced innovative travel insurance products tailored to the pandemic challenges. With the swift digitalization of the economies during a pandemic, opportunities for the growth of Fintech have expanded. Fintech has played a significant role in enhancing the tourism experience in the Asian region through various technological interventions, viz. facilitating international transactions, convenient online booking and reservations, currency exchange, and accessing peer-to-peer services. Mobile payment apps like Alipay and WeChat Pay are widely accepted in many Asian countries, making it easy for the tourist to make transactions. These apps have also enabled currency conversion and seamless international transactions, reducing the need for physical cash and foreign currency exchange. Embracing Fintech opportunities and implementing policies that facilitate safe financial innovation and adoption can significantly contribute to boosting tourism and, consequently, fostering economic development in a nation.

How Asia Used Fintech to Address Tourism Challenges Posed by the Pandemic—Case Studies of Successful Fintech Applications in Asia's Tourism Sector.

India's Fintech industry is often hailed as a global exemplar of financial innovation (Ozili et al., 2024), with results which are pertinent and applicable for all economies, regardless of their development stage (D'Silva et al., 2019; Migozzi et al., 2023). India, a traditionally cash-driven economy has responded well to the fintech opportunity, mainly triggered by a surge in e-commerce, and Smartphone penetration and is responsible for bringing a large number of Indians into the formal economy (Mashal & Kumar, 2023). It is one of the world's biggest emerging hotspots for Fintech and emerged as one of the quickly emerging digital economies ("India has the largest Fintech ecosystem in the world after US, China: Study," 2022; Aggarwal et al., 2023). In India, the growth of Fintech is also facilitated by the strong presence of the tech workforce. The country has leapfrogged credit card-based payments with digital payments. India has taken a few steps that have made the leapfrog possible, including bringing more people into the formal financial system through easy procedures of opening bank accounts and various government initiatives such as Jan Dhan Yojana, Aadhar, and the emergence of Unified Payment Interface (UPI) (CAPAM2018, Blueprint for Capital Market in New India 2011, 2018).

India's journey in redefining its digital financial landscape reached a significant milestone with the introduction of the UPI (Unified Payments Interface) system. By simplifying and streamlining banking transactions, UPI has removed the traditional barriers associated with interbank transfers. Tourists, both domestic and international, no longer have to grapple with the complexities of various bank interfaces or endure prolonged wait times for transaction approvals. Instead, UPI's real-time interbank transaction capability ensures swift, secure, and efficient financial interactions, regardless of the bank's origin or destination. This has not only made India's financial ecosystem more inclusive but has also amplified its attractiveness as a tourist destination where financial transactions are hassle-free. Moreover, government initiatives such as "Make in India'" and "Digital India" have played a pivotal role in accelerating the adoption of Fintech. The demonetization of high-value banknotes in 2016 provided a boost to the sector by increasing awareness and adoption of Fintech ("KPMG report: Fintech in India—Powering mobile payment," 2019). As India embarks on a significant digital transformation journey towards a cashless society, one of the world's foremost business leaders has expressed admiration for the efforts of the government led by Prime Minister Narendra Modi in fostering a digital payment ecosystem. During his address to the media at the Dell Technologies World 2018, Michael Dell, Chairman and CEO of Dell Technologies remarked

> *IndiaStack, for example, is a world-leading concept and a techie, I get very inspired by what India is doing. It is an amazing government-led effort,*
> "Modi's digital India has a new admirer," 2018.

A notable behavioural shift is evident as societies move away from cash-based economies and increasingly adopt user-friendly digital payments. Today, digital transactions are commonplace in India, even for modest amounts like a 10-cent cup of milk tea or a $2 purchase of fresh vegetables (Mashal & Kumar, 2023).

"The little QR code is ubiquitous across India's vastness. You find it pasted on a tree next to a roadside barber, propped on the pile of embroidery sold by female weavers, sticking out of a mound of freshly roasted peanuts on a snack cart. A beachside performer in Mumbai places it on his donations can before beginning his robot act; a Delhi beggar flashes it through your car's window when you plead that you have no cash" (Mashal & Kumar, 2023). In India, the Reserve Bank of India (RBI) has been instrumental in enabling the development of the fintech sector. They have adopted a prudent stance in handling issues related to consumer protection and law enforcement. They focus on establishing a conducive environment facilitating fintech innovations, extending the accessibility of banking services to the unbanked demographic, overseeing digital payment, and offering choices to consumers. Fintech is shaping the tourist industry in India by offering convenient and secure payment options, cashback incentives, and streamlining financial transactions. The

integration of Fintech into the tourism sector is making travel more accessible and enjoyable for both domestic and international tourists. The increasing adoption of these trends is positioning India as an attractive market worldwide and having a very positive impact on the tourism sector.

China's digital metamorphosis in recent years, especially in its payment landscape, stands as a testament to the potent synergy of technology and commerce. Platforms like Alipay and WeChat Pay have not only revolutionized the way transactions occur but have also redefined the consumer experience. These aren't just payment systems; they are comprehensive ecosystems. Tourists in China, whether domestic or international, benefit from the unparalleled convenience these platforms offer. From booking accommodations, purchasing tickets to attractions, availing transportation services, to leaving reviews everything can be managed through a singular, integrated application. This holistic approach minimizes friction, creating a seamless journey for consumers, while businesses, irrespective of size, gain access to a vast user base and invaluable data analytics.

The Smart Nation effort in Singapore has established a noteworthy standard on a worldwide scale for the successful use of technology in several aspects of everyday life. This progressive initiative encompasses more than a mere assemblage of digital solutions; it represents a comprehensive strategy aimed at fostering a society that flourishes via innovation, effectiveness, and interconnectedness. One notable aspect is its significant influence on the tourism industry. When individuals visit Singapore, they are provided with an exceptional digital encounter. This is facilitated by the implementation of cashless payment systems, which guarantee seamless transactions across various services. Additionally, the presence of digital kiosks enables immediate access to information and assistance. Furthermore, the incorporation of intelligent transportation solutions, such as automated metros and e-scooters, offers commuters convenient options for travel. The integration of digital immersion not only enhances the overall visitor experience but also establishes Singapore as a forward-looking vacation destination.

12.3.8 Analysis of Countries in Asia That Have Successfully Utilized Fintech and How Africa Can Draw Insights

Drawing from India's success story of Unified Payments Interface, Africa, with its mosaic of nations and banking systems, stands to gain immensely from a similar unified approach. A pan-African payment interface could revolutionize the way tourists experience the continent. Instead of navigating a labyrinth of distinct banking protocols in every country, a unified system would allow for smoother cross-border financial interactions. Such an initiative would not only enhance the tourist experience but also foster regional collaboration and integration among African nations. As Africa continues to position itself as a prime tourist destination, adopting a UPI-inspired model could be a strategic move to enhance its economic and tourist appeal on the global stage. Drawing inspiration from China's digital revolution in Payment Systems, Africa, with its burgeoning tourism industry and growing digital infrastructure, finds itself at a precipice of opportunity. By developing or adopting comprehensive digital platforms tailored to its unique challenges and opportunities, Africa can foster a more interconnected and user-friendly tourism ecosystem. Such platforms could integrate local cultural experiences, safaris, historical site bookings, and more, offering tourists a one-stop solution while democratizing access for local businesses. Embracing this digital shift, inspired by the Chinese model, could further position Africa as a modern and tech-savvy tourist destination while ensuring benefits permeate through every layer of its tourism industry.

Singapore's smart Nation vision presents tremendous lessons for African nations. The indisputable diversity and richness of Africa's natural and cultural attractions can be further enhanced by incorporating technology to augment the visitor experience, adding substantial value. By conducting a comprehensive analysis and incorporating key components from Singapore's Smart Nation plan, African nations have the potential to revolutionize their prominent tourist centres into technologically advanced destinations. This statement does not suggest a simple duplication but rather emphasizes the need for a deliberate incorporation of technology that aligns with and enhances the distinct characteristics of the African setting. One example of how digital technologies can enhance efficiency is by facilitating the reservation process for safaris, heritage sites, and cultural events. The implementation of intelligent transportation systems can enhance the accessibility of isolated attractions. In essence, through the fusion of Africa's innate aesthetic allure and a measure of technical advancement, the continent has the potential to redefine its attractiveness to the international tourism sector. This redefinition would entail providing visitors with genuine cultural encounters while also giving contemporary amenities and comforts.

For the development of Fintech, the economy needs to have the availability of necessary technical skills and expertise, technological readiness of the market and introduction of supportive government policies, favourable regulatory environment ("KPMG report: Fintech in India Powering mobile payments," 2019). The regulatory barriers can prevent and slow down the adoption of technology-based financial products in emerging markets. To support the deepening of financial innovation and adoption, modernized and updated financial infrastructure, strong cross-border coordination and sharing of information is required (Feyen et al., 2023). Fintech-enabled business models and products can support economies to become more resilient and promote an equitable recovery from the pandemic (Demirgüç-Kunt et al., 2022). The digitalization of financial services has the potential to significantly boost a country's tourism industry, thereby contributing to its economic development. As such, nations should embrace Fintech opportunities and establish policies that facilitate and promote secure financial innovation and adoption as Fintech has the potential to play a crucial role in promoting tourism by enhancing the overall travel experience, making transactions more convenient and secure, and boosting growth of the tourism industry.

12.4 Conclusion and Policy Recommendations

In conclusion, the global tourism industry bore the brunt of the COVID-19 pandemic, witnessing a stark decline in airport activity, deserted tourist hotspots, and growing uncertainty among businesses. However, the tourism sector in Asia has staged a remarkable comeback, thanks to the widespread distribution of vaccines and the relaxation of travel restrictions, largely driven by advancements in Financial Technology (Fintech). Innovations such as touchless payment systems, digital identity verification, and blockchain-based travel documentation have become instrumental in meeting the evolving health and safety expectations of travellers. This chapter has delved into the strategic utilization of Fintech in Asia as a catalyst for the revival of the tourism industry post-pandemic. Furthermore, it has illuminated potential ramifications and opportunities that Africa may encounter as it navigates this new landscape. Embracing the adoption of Financial Technology (Fintech) is not just desirable but imperative for Africa, as it seeks not only to recapture its former prominence in the tourism sector but also to chart a progressive and forward-looking course for its tourism industry. In this ever-evolving global tourism ecosystem, Fintech emerges as a powerful tool that holds the promise of not only rejuvenating the industry but also shaping its future in a way that is inclusive, innovative, and resilient. It underscores the dynamic nature of the sector and the endless possibilities for adaptation and growth in the face of challenges. Africa, like Asia, has the potential to harness the transformative potential of Fintech to redefine its role in the global tourism landscape, offering both travellers and local businesses a vibrant and promising future.

References

Adeleye, B. N., Ogede, J. S., Rabbani, M. R., Adam, L. S., & Mazhar, M. (2022a). Moderation analysis of exchange rate, tourism and economic growth in Asia. *PLoS One, 17*(12), e0279937.

Adeleye, B. N., Adam, L. S., Ahmad, P., & Ola-David, O. (2022b). Investigating tourism and exchange rate dynamics on economic growth in Sri Lanka. *Journal of Policy Research in Tourism, Leisure and Events*, 1–17.

Aggarwal, M., Nayak, K. M., & Bhatt, V. (2023). Examining the factors influencing fintech adoption behaviour of gen Y in India. *Cogent Economics & Finance, 11*(1). https://doi.org/10.1080/23322039.2023.2197699

Amin, S. B., & Taghizadeh-Hesary, F. (2023). Tourism, sustainability, and the economy in Bangladesh: The innovation connection amidst Covid-19. *Economic Analysis and Policy*.

Awotunde, J. B., Adeniyi, E. A., Ogundokun, R. O., & Ayo, F. E. (2021). Application of big data with Fintech in financial services. In P. M. S. Choi & S. H. Huang (Eds.), *Fintech with artificial intelligence, big data, and blockchain. Blockchain technologies*. Springer. https://doi.org/10.1007/978-981-33-6137-9_3

Baralla, G., Pinna, A., Tonelli, R., Marchesi, M., & Ibba, S. (2021). Ensuring transparency and traceability of food local products: A blockchain application to a Smart Tourism Region. *Concurrency and Computation: Practice and Experience, 33*(1), e5857.

Benaraba, C. M. D., Bulaon, N. J. B., Escosio, S. M. D., Narvaez, A. H. G., Suinan, A. N. A., & Roma, M. N. (2022). A comparative analysis on the career perceptions of tourism management students before and during the COVID-19 pandemic. *Journal of Hospitality, Leisure, Sport & Tourism Education, 30*, 100361.

Buhalis, D., Harwood, T., Bogicevic, V., Viglia, G., Beldona, S., & Hofacker, C. (2019). Technological disruptions in services: Lessons from tourism and hospitality. *Journal of Service Management, 30*(4), 484–506.

Büscher, B., & Fletcher, R. (2019). Towards convivial conservation. *Conservation and Society, 17*(3), 283–296.

CAPAM2018 'Blueprint for Capital Market in New India 2022' The Experts' Voice – A compendium of articles: Foreword. (2018). *FICCI Studies and Surveys*.

Cong, L. W., & He, Z. (2019). Blockchain disruption and smart contracts. *The Review of Financial Studies, 32*(5), 1754–1797. https://doi.org/10.1093/rfs/hhz007

D'Silva, D., Filková, Z., Packer, F., & Tiwari, S. (2019). The design of digital financial infrastructure: Lessons from India. *BIS Paper, 106*.

Demirgüç-Kunt, A., Klapper, L., Singer, D., Ansar, S., & Hess, J. (2018). *The global findex database 2017* (1st ed.). World Bank Publications. https://doi.org/10.1596/978-1-4648-1259-0

Demirgüç-Kunt, A., Klapper, L., Singer, D., & Ansar, S. (2022). The global findex database 2021: Financial inclusion, digital payments, and resilience in the age of COVID-19. World Bank Publications.

Dwivedi, Y. K., Kshetri, N., Hughes, L., Slade, E. L., Jeyaraj, A., Kar, A. K., et al. (2023). "So what if ChatGPT wrote it?" Multidisciplinary perspectives on opportunities, challenges and implications of generative conversational AI for research, practice and policy. *International Journal of Information Management, 71*, 102642.

Feyen, E., Natarajan, H., Saal, M. I., & ProQuest. (2023). *Fintech and the future of finance: Market and policy implications*. World Bank Group.

Five key fintech trends to watch in 2022 according to Buckzy Payments: In the year ahead, consumer behavior and early adoption of new technology are set to transform the financial services market. Banks need to adapt. (2021, November 18). *PR Newswire*. https://www.proquest.com/wire-feeds/five-key-fintech-trends-watch-2022-according/docview/2598431443/se-2

Gössling, S., Scott, D., & Hall, C. M. (2020). Pandemics, tourism and global change: A rapid assessment of COVID-19. *Journal of Sustainable Tourism, 29*(1), 1–20.

Goldstein, I., Jiang, W., & Karolyi, G. A. (2019). To FinTech and beyond. *The Review of Financial Studies, 32*(5), 1647–1661.

Higgins-Desbiolles, F. (2022). The ongoingness of imperialism: The problem of tourism dependency and the promise of radical equality. *Annals of Tourism Research, 94*, 103382.

India has the largest FinTech ecosystem in the world after US, China: Study. (2022). In TECHCiRCLE. HT Digital Streams Limited.

Kaurav, R. P. S., Baber, R., & Rajput, S. (2020). Technology-driven tourism and hospitality industry as a tool for economic development: A bibliometric analysis. In *The emerald handbook of ICT in tourism and hospitality* (pp. 469–486). Emerald Publishing Limited.

Khan, M. A. A., & Hashim, H. (2020). The effect of COVID-19 on tourism and hospitality industry in Malaysia, resurgence in the post-pandemic era: A conceptual criterion. *International Journal of Tourism & Hospitality Review, 7*(2), 54–62.

Khayer, A., & Alam, S. (2022, October 21). Application of management information systems in the financial sector: An overview of fintech innovations. In R. Hoque (Ed.), *Management information systems and digitalization: Current status and future trends* (pp. 85–125).

KPMG report: Fintech in India – Powering mobile payments launches at Nasscom India Fintech Day. (2019). In United News of India. HT Digital Streams Limited.

Leung, W. K., Chang, M. K., Cheung, M. L., & Shi, S. (2023). VR tourism experiences and tourist behaviour intention in COVID-19: An experience economy and mood management perspective. *Information Technology & People, 36*(3), 1095–1125.

Lontchi, C. B., Yang, B., & Shuaib, K. M. (2023). Effect of financial technology on SMEs performance in Cameroon amid COVID-19 recovery: The mediating effect of financial literacy. *Sustainability, 15*(3), 2171.

Mashal, M., & Kumar, H.. (2023). Where digital payments, even for a 10-cent chai, are colossal in scale. *New York Times* (Online).

Mhlanga, D. (2020). Industry 4.0 in finance: The impact of artificial intelligence (AI) on digital financial inclusion. *International Journal of Financial Studies, 8*(3), 45.

Mhlanga, D. (2021). Factors that matter for financial inclusion: Evidence from sub-Saharan Africa-The Zimbabwe Case. *Academic Journal of Interdisciplinary Studies, 10*(6), 48–48.

Mhlanga, D. (2022a). Selected digital financial inclusion success stories across developing economies. In *Digital financial inclusion: Revisiting poverty theories in the context of the fourth industrial revolution* (pp. 301–322). Springer International Publishing.

Mhlanga, D. (2022b). COVID-19 and digital financial inclusion: Policies and innovation that can accelerate financial inclusion in a post-COVID world through fintech. *African Journal of Development Studies, 2022*(si2), 79.

Mhlanga, D. (2023a). *Artificial intelligence in elderly care: Navigating ethical and responsible AI adoption for seniors*. Available at SSRN 4675564.

Mhlanga, D. (2023b). The role of FinTech and AI in agriculture, towards eradicating hunger and ensuring food security. In *FinTech and artificial intelligence for sustainable development. Sustainable development goals series*. Palgrave Macmillan. https://doi.org/10.1007/978-3-031-37776-1_6

Migozzi, J., Urban, M., & Wójcik, D. (2023). "You should do what India does": FinTech ecosystems in India reshaping the geography of finance. *Geoforum*, 103720. https://doi.org/10.1016/j.geoforum.2023.103720

Modi's digital India has a new admirer: What Michael Dell has to say about India's digital payments drive; As India looks to take big digital leap into moving towards a cash less society, one of the world's top business leaders has lauded Prime Minister Narendra Modi-led government's efforts to promote a digital payments ecosystem. (2018). AutoParts Asia.

Morgan, P. J. (2022). Fintech and financial inclusion in Southeast Asia and India. *Asian Economic Policy Review, 17*(2), 183–208. https://doi.org/10.1111/aepr.12379

Ozili, P. K., Mhlanga, D., Ammar, R., & Fersi, M. (2024). Information effect of fintech and digital finance on financial inclusion during the COVID-19 pandemic: Global evidence. *FinTech, 3*(1), 66–82.

Park, S., Kim, Y. R., & Ho, C. S. T. (2022). Analysis of travel mobility under COVID-19: Application of network science. *Journal of Travel & Tourism Marketing, 39*(3), 335–352.

Purba, J. H. V., Fathiah, R., & Steven, S. (2021). The impact of Covid-19 pandemic on the tourism sector in Indonesia. *Riset: Jurnal Aplikasi Ekonomi Akuntansi dan Bisnis, 3*(1), 389–401.

Puschmann, T. (2017). Fintech. *Business & Information Systems Engineering, 59*, 69–76.

Rabbani, M. R., Abdulla, Y., Basahr, A., Khan, S., & Moh'd Ali, M. A. (2020, November). Embracing of Fintech in Islamic finance in the post COVID era. In *2020 international conference on decision aid sciences and application (DASA)* (pp. 1230–1234). IEEE.

Sandelowski, M., & Barroso, J. (2006). *Handbook for synthesizing qualitative research*. Springer Publishing Company.

Seyfi, S., Hall, C. M., & Shabani, B. (2023). COVID-19 and international travel restrictions: The geopolitics of health and tourism. *Tourism Geographies, 25*(1), 357–373.

Škare, M., Soriano, D. R., & Porada-Rochoń, M. (2021). Impact of COVID-19 on the travel and tourism industry. *Technological Forecasting and Social Change, 163*, 120469.

Suyunchaliyeva, M. M., Nautiyal, R., Shaikh, A. A., & Sharma, R. (2021). The use of mobile payment systems in post-COVID-19 economic recovery: Primary research on an emerging market for experience goods. *Sustainability, 13*(24), 13511.

Teichmann, F. M. J., & Falker, M. C. (2021). Cryptocurrencies and financial crime: Solutions from Liechtenstein. *Journal of Money Laundering Control, 24*(4), 775–788.

Toumi, A., Najaf, K., Dhiaf, M. M., Li, N. S., & Kanagasabapathy, S. (2023). The role of Fintech firms' sustainability during the COVID-19 period. *Environmental Science and Pollution Research International, 30*(20), 58855–58865. https://doi.org/10.1007/s11356-023-26530-3
Vena-Oya, J., Castañeda-García, J. A., & Rodríguez-Molina, M. Á. (2022). Forecasting a post-COVID-19 economic crisis using fuzzy cognitive maps: A Spanish tourism-sector perspective. *Current Issues in Tourism, 25*(13), 2048–2062.

Chapter 13
Conclusion: Tourism, Digitalisation, and Sustainable Development

Emmanuel Ndhlovu, Kaitano Dube, and Ngoni Courage Shereni

13.1 Introduction

Digitalisation significantly transforms business systems, processes and markets, leveraging digital technologies (Dredge et al., 2019). In this regard, the contemporary tourism and hospitality industry is now moving towards being a fully digitalised economic sector. Undoubtedly, the COVID-19 pandemic accelerated the digitalisation being experienced in today's tourism industry and brought about changes never imagined before. However, it is important to note that the tourism and hospitality industry has been infusing digital technologies for decades even before the COVID-19 crisis (Abouzeid, 2022). The airline industry is acknowledged as one of the early adopters of digital technology in the form of Global Distribution Systems (GDS). Similarly, travel agents became important in the aviation value chain because they could integrate their operations with airline booking platforms. In the same vein, the hospitality industry is also regarded as a pioneer in the digitalisation of the tourism industry by adopting computer reservation systems and property management systems. That being said, it is clear that the tourism and hospitality industry has always embraced digitalisation before COVID-19.

Currently, the fourth industrial revolution is an important development in digitalisation, not only in the tourism and hospitality industry but also in many different economic sectors around the globe. The adoption of the fourth industrial revolution in the tourism industry gave rise to what is now known as Tourism 4.0. This era is characterised by disruptive technologies that challenge the conventional way of

E. Ndhlovu · K. Dube
Vaal University of Technology, Vanderbijlpark, South Africa
e-mail: kaitanod@vut.ac.za

N. C. Shereni (✉)
School of Tourism and Hospitality, College of Business and Economics, University of Johannesburg, Johannesburg, South Africa

E. Ndhlovu et al. (eds.), *Tourism and Hospitality for Sustainable Development*,
https://doi.org/10.1007/978-3-031-63069-9_13

doing things. Big data analytics, the Internet of Things, AI, robotics, cloud computing, blockchains, virtual reality, and augmented reality are some of the fourth industrialisation digital tools changing the face of the tourism industry (Rodrigues et al., 2023). Tied to this is the growth of user-generated content due to increasing online customer interactions. Today's customers have a high digital footprint and leave large trails of data that technologically savvy enterprises can utilise to predict the future behaviour of their clients (Shereni & Chambwe, 2020). These days customers can easily make airline bookings, hotel reservations, online payment, and destination choices over digital platforms (Hoang-tien et al., 2024). The emergence of disruptive technologies, therefore, makes it easy to generate, process and store such user-generated data to inform future business-related decisions.

It is without doubt that for establishments in the tourism and hospitality industry to remain competitive and to grow sustainably they need to move with changes in technology. More so, the exponential growth in smart technologies has increased connectivity at an individual and business level. It is now easy for businesses to reach out to customers through various digital platforms such as social media and mobile applications. As such, tourism businesses need to consolidate their online presence to be visible to both individual and corporate customers. The chapters in this volume provide a link between the tourism and hospitality industry, sustainable development and digitalisation. The main emphasis is the growing influence of digitalisation in the tourism and hospitality industry and its contribution to sustainable development.

This chapter provides concluding remarks for this volume. In essence, it summarises contributions made by different authors and provides a future research agenda emanating from various contributions. The chapter is organised as follows: a discussion of digitalisation and sustainable development; key findings and conclusions, recommendations for industry practitioners and policymakers, and future research agenda.

13.2 Digitalisation and Sustainable Development

As the world is increasingly focusing on sustainability, digitalisation is now being seen as a critical tool in enhancing sustainable tourism development (El Archi et al., 2023). With a few years to go before the era of Sustainable Development Goals (SDGs) comes to an end, it becomes important to focus attention on the contribution of the tourism and hospitality industry to the Agenda 2030. The COVID-19 pandemic has shown a greater need to align business and social processes with the SDGs (Rodrigues et al., 2023). It is, therefore, suffice to say digitalisation can help to restore the momentum lost in the drive towards the achievement of the SDGs due to the disruptions caused by the COVID-19 pandemic. However, little is known about the potential of digitalisation in sustainable tourism development especially in the post-COVID-19 era. This volume, therefore, utilised various tourism and hospitality case studies from different global contexts to bring to the fore how digital solutions can play a vital role in improving business processes and contribute to sustainable tourism development.

The intersection of digitalisation and sustainable tourism provides opportunities for new product development and improved viability of tourist destinations. E-tourism and smart tourism are some of the concepts that have emerged out of the intersection between tourism and digitalisation. These concepts have gone a long way in helping destinations grow sustainably by optimising resource use and minimising the negative impacts of tourism. Digitalisation is used to promote sustainable tourism in a number of ways. Tourism and hospitality organisations can leverage the growth of social media to raise awareness of sustainability issues, encourage tourists to adopt sustainable behaviours and engage effectively with local communities (El Archi et al., 2023). There is also a growing realisation of the importance of social media and corporate websites in reporting on sustainability and climate change issues (Shereni, 2023; Shereni et al., 2022). This is done to increase transparency and to ensure that organisations gain the trust of key stakeholders.

Evidence from literature suggests that smart tourism supports the development of sustainable business models (Rodrigues et al., 2023). Smart tourism components such as virtual reality and artificial intelligence enable tourists to access destinations remotely thereby reducing carbon emissions significantly (Abouzeid, 2022). In addition, the use of motion sensors and other technologies in energy management points to the central role played by smart tourism in sustainable development and climate change mitigation. Augmented reality has been used often to provide more information about tourist attractions in a way that enhances tourist experiences and promotes local participation. The concept of smart cities has also emerged strongly in the last decade more so with the drive to achieve SDG 11 on sustainable cities and communities. Smart cities integrate information communication technologies in urban destinations to ensure seamless processes and efficient use of resources as well as to enhance service provision (Woyo & Ukpabi, 2022). Studies on smart cities are much concentrated in developed countries suggesting a wider adoption of the concept by the Global North as compared to countries in the Global South with Africa in particular still lagging.

Digitalisation also plays a vital role in enhancing the economic sustainability dimension. Economic sustainability is anchored on among other things the promotion of a circular economy and developing a resilient economic system that is innovative, open, and competitive (Renn et al., 2021). The role of digitalisation in the circular economy is to develop processes that ensure waste reduction and foster the idea that waste from one economic system is used as input in another. Within the tourism and hospitality industry digital innovations such as the sharing economy, business models minimising waste and promoting recycling are key features of the circular economy (Sørensen & Baerenholdt, 2020). Also, one of the critical arguments that emerged from this volume is the role of digitalisation in building a resilient tourism sector and viable businesses that can exist in the future. This resonates with the tenets of economic sustainability.

13.3 Key Findings and Conclusions

This volume explored four themes which are: conceptualisation and review of digital revolution in tourism; digital innovation in tourism and hospitality; barriers to and challenges to technology adoption in tourism businesses; and policy and practical recommendations. This section provides key contributions from this volume in line with the themes discussed.

The first section of this volume, constituting one chapter, conceptualises digitalisation in the tourism and hospitality sector. .

Tourism and hospitality training institutions, especially those in sub-Saharan Africa, have recently started infusing digital technologies at an increasing rate. The COVID-19 pandemic exposed the weaknesses of fully depending on face-to-face learning models and raised the need for digital transformation in tourism and hospitality training. In Chap. 2, Makuyana, Ndhlovu and Dube use the Multiple Intelligence Theory (MIT) to argue for the digitalisation of tourism and hospitality education in a way that prepares learners for the digitalised work environment. The authors maintain that even though the use of MIT is low in tourism and hospitality education in sub-Saharan Africa there is evidence of integration of technologies like e-commerce, e-learning, and virtual reality. The chapter concludes that the integration of digital technologies varies across geographical contexts due to differences in the demand exerted by the labour markets of a particular country, the structure of the tourism curricula and the competencies of course instructors. The study emphasised the need to infuse MIT in digital learning to cater for the different needs of learners and to develop multiple sets of skills amongst tourism and hospitality students.

The second section examined various digital innovations in the tourism and hospitality industry. As noted in the previous section digital transformation in the industry has been achieved mostly by adopting cutting-edge innovations. Chapters in this section identify and discuss such digital innovations in various sub-sectors of the tourism and hospitality industry. In Chap. 3, Chihwai provides a glimpse into the digital innovations adopted by South African national parks, hotels and airports. The study observes that South African airports are making strides to adopt various digital technologies. Although there are inadequacies in adopting innovative technologies, there are plans to adopt self-service check-in, self-boarding e-gates and baggage drops. In South African national parks, digital innovation is utilised to a lesser extent. The South African National Parks Management (SANParks) deploys a digital tool called Management Effectiveness Tracking Tool (METT), mostly in conservation efforts. It also uses various social media tools to carry out campaigns, reach out to customers, and promote various national park offerings. The study found that hotels adopt digital innovations in property management, reservations, customer rewards programmes, and procurement. The study concludes that the adoption of digital innovations at airports, national parks and hotels results in improved customer satisfaction, an increase in profits and improvements in operational efficiency.

In Chap. 4, Moyo, Zimusi and Ndhlovu established that women-owned restaurants in Zimbabwe use digital technologies such as social media, mobile payment platforms and point-of-sale services. There is a varied application of digital innovation in the establishments under study depending on their size, with some going to the extent of using digital menus, procurement and stock management technologies. The study concluded that various factors drive women-owned restaurants to adopt digital innovations. Such factors include perceived usefulness, perceived ease of use, and attitude towards use. The study concluded that the digital technologies adopted help restaurant owners improve customer service, giving them a competitive advantage. Furthermore, the technology used was deemed easy to use, which has a bearing on the attitude of female restaurant owners towards adopting digital technologies. Shereni and Ndhlovu (Chap. 5) revealed that Zimbabwe has a vibrant urban tourism product that can be used to spur growth in accommodation peer-to-peer business. Regardless of the potential P2P accommodation platforms have in Zimbabwe's urban tourism landscape, their adoption is still limited. Challenges that hinder the growth of this business model include the unstable macroeconomic environment, the unpredictable political environment and economic sanctions stifling business growth. That being said, the significant informal economy in the urban areas of Zimbabwe provides entrepreneurial opportunities for residents to monetise their assets through P2P platforms and even conceptualise homegrown P2P business models. The global aviation industry is constantly adopting novel digital innovations. In Chap. 6, Ndhlovu, Makuyana, and Dube established the state of hi-tech digital innovations using case studies from various global airlines. COVID-19 is a major driver of the digital explosion occurring in the aviation industry. The study revealed that many of the hi-tech innovations by airlines globally are in the domain of AI, AR, VR, big data analytics, blockchains, cloud computing, robotics, and the Internet of Things. The chapter concludes that the major motivation for airline digitalisation is improved operational efficiency, providing predictive and preventive solutions, flow monitoring and capacity management, and reduced operational costs.

The focus of the third section of this volume is the barriers to and challenges of technology adoption in tourism businesses. Chapters in this section discuss the challenges hindering digital technology adoption in various tourism and hospitality subsectors. For example, in Chap. 7, Zhou and Chibaya explored the opportunities and challenges of using digital technologies in wildlife conservation in Zimbabwe. Wildlife conservation faces threats from poaching, wildfires and the effects of climate change. Technological innovations can be adopted to ameliorate such threats and aid conservation efforts. The chapter argues that digitalisation can be harnessed in wildlife conservation through the use of CCTV cameras, unmanned aerial vehicles (UAVs), drones, mobile technologies, GPS, satellite mapping, and Radio Frequency Identification (RFID) tags. However, numerous challenges are faced when employing the identified technological innovations. The challenges include budgetary constraints, skills shortage, and vandalism of infrastructure and network challenges.

In Chap. 8, Makuyana, Ndhlovu, and Dube highlighted that the diversity of impairments among people with disability and their different digital needs makes it difficult for restaurants to create inclusive service encounters. The chapter concludes that restaurants in sub-Saharan Africa still lag in implementing digitalisation that considers people with disability. Considering that people with disability represent a significant market in the tourism and hospitality industry, it will be prudent to coopt inclusive digital innovations in the restaurant business. The developments in the digital economy have also presented challenges to the travel agent business. The role of Online Travel Agents (OTA) in the tourism value chain has increasingly grown in importance of late. This has seen sub-sectors such as accommodation having to rely more on OTA amidst dwindling numbers booking directly with them. This comes at a cost to hotels, such as paying commissions to third parties and losing direct access to their customers. Nyathela-Sunday and Banoobhai-Anwar (Chap. 9) provided insights into the challenges faced by hotels in Cape Town, South Africa, in an attempt to get direct bookings. The authors indicated challenges such as high acquisition costs for direct bookings and guests preferring to book with OTA rather than directly with relevant hotels and hotels using websites managed by third parties.

Chapters in the fourth and final section of this volume address policy and practical recommendations. In Chap. 10, Ndhlovu, Dube, and Shereni examine the strategies used by small hospitality and tourism enterprises to promote inclusive technological adoption and integration. The authors identified factors affecting the adoption of digitalisation in small tourism and hospitality enterprises, such as financial barriers, education and training, infrastructural challenges, organisational culture, and social barriers. The findings recommend that for digitalisation to be inclusive, there is a need to provide such enterprises with financial support as well as educate and train them in digital skills to help change their attitude towards embracing digital technologies in their organisations. Ndhlovu and Montsiemang (Chap. 11) sought to identify technology adoption challenges in Africa and proffer solutions to these challenges. There is a notable dearth of literature focusing on technology adoption and integration in the tourism industry in Africa. This is the gap that this chapter aimed to close with the view first to understand the nature of these challenges. The challenges identified in this study include a lack of financial capacity to keep pace with changes in technology, a skills deficit, employees' unwillingness to adopt novel digital solutions, and data protection and security issues. Further to providing a preview of the challenges, the chapter proposed interventions to skirt around the various concerns identified. Allocating enough financial resources towards acquiring relevant technological infrastructures needed to foster a culture of innovation is regarded as an essential measure. In addition, training, collaboration with other players, engaging with industry associations, staying abreast of the latest technological advancements, and placing a premium on technologies enhancing customer experience are equally critical digital solutions. Lastly, Mhlanga and Sethi explored financial technology (Fintech) innovations in Asia for post-COVID-19 tourism recovery and critical lessons for Africa. The study observes that Fintech combines finance and technology to provide digital financial services to the business world. Fintech has revolutionised the financial sector, changing how payments, insurance and investment decisions are made across different economic sectors. In Asia, Fintech is recognised for enhancing the tourist experience by facilitating international transactions, making convenient online bookings and reservations, facilitating currency exchange, and providing access to peer-to-peer services. Experiences from Asia show that traditional cash economies were forced to adopt Fintech solutions due to the dictates of the COVID-19 pandemic, and this trend has continued in the post-COVID-19 era. The mushrooming of online payment platforms across Asia has immensely benefited the tourism industry, and Africa can draw lessons from such developments. The authors call for pan-African digital payment platforms that simplify the banking system in a way that enhances tourist experiences. Innovations such as touchless payment systems, digital identity verification, and blockchain-based travel documentation played a critical role in the recovery of the Asian tourism industry, providing African economies with an opportunity to earn significant tourism revenues and boost the tourism industry post-COVID-19.

13.4 Recommendations for Industry Practitioners and Policymakers

This section compiles the implications and recommendations for industry practitioners and policymakers that emerged from the contribution in this volume. Digitalisation will continue to play a significant role in sustainable tourism development. The changes brought about by the COVID-19 pandemic compel tourism and hospitality businesses to adopt various technological innovations that allow them to be competitive and make them resilient in the face of future pandemics.

For tourism and hospitality businesses to benefit from the digital explosion currently being spearheaded by the fourth industrial revolution, there is a need to invest in the proper technological infrastructure that is compatible with the organisational size and existing technological systems. Furthermore, management support and commitment are essential aspects of creating a conducive climate for technological innovation. Leung et al. (2011) emphasise the role played by management support in adopting digital technological innovations. It is also essential for the tourism and hospitality industry to lobby the government and policymakers to implement laws that promote digital innovations. Laws and regulations that constrict digital innovation are detrimental to the digitalisation of the tourism and hospitality industry (Baker, 2012). On a different note, considering the ever-changing technological environment, it is critical for learning institutions to continuously change their tourism curricula to keep abreast with the evolving technological environment. The modern tourism and hospitality environment is characterised by different technological innovations such as robotics, artificial intelligence, and virtual reality, among others (Tung & Au, 2018). It, therefore, becomes critical for learning institutions to integrate digitalisation into the learning process to produce graduates who are fit for the industry.

The contributions from this volume demonstrated the necessity to build a team of competent technicians who can navigate the complex digital terrain of tourism and hospitality. Digital innovations enable businesses in the tourism and hospitality sector to improve the quality of products, lower costs, increase efficiency, meet customers' changes, increase sales and profits, gain market share and have a competitive advantage (Ndhlovu & Dube, 2023a, b). As such, having a solid team working on providing digital solutions will aid companies in the tourism and hospitality industry enjoy these advantages. At a destination and country level, it would be prudent to come up with home-grown digital solutions. Examples cited in this volume include coming up with Fintech innovations grounded in the African context for the African market and also encouraging local entrepreneurs to come up with innovations such as peer-to-peer platforms serving the local market.

Nature-based tourism is significant in many countries in Africa, and it generates millions in revenue (Zhou, 2023). This makes wildlife conservation critical to maintain this market's viability and ensure the preservation of wildlife resources. Digitalisation has been successfully implemented in conservation efforts in many countries. Authorities managing wildlife in protected areas must integrate sophisticated digital tools to counter the threats posed by poachers and natural occurrences such as climate change. Furthermore, this volume has shown that inclusive digitalisation, considering people with disabilities, is still very limited in the tourism and hospitality industry, especially in developing economies. However, studies have shown that inclusive digitalisation empowers people with disabilities empowers. It is, therefore, vital for the tourism and hospitality industry to introduce digital tools such as voice-activated chatbots, service robots, digital room keys, and audio descriptions of service offerings.

13.5 Future Research Agenda

This volume contributes to global discussions on the nexus between digitalisation, sustainable development and the tourism and hospitality industry. The book set out various research agendas that can be pursued in future studies. Firstly, numerous challenges the tourism and hospitality sector encountered in adopting digital innovations, especially in the Global South, were unearthed. There is a need to conduct research in various industry sub-sectors to gain a deeper appreciation of these challenges and proffer informed, practical solutions. In addition, the link between tourism education and industry practices is obscured in the literature. Properly designed curricula can tailor industry digital needs by producing employees with the requisite skills. Future studies can explore this gap by looking at the industry needs vis a vis what training institutions are producing. Also, there is a notable gap between digital innovations in the Global South and the Global North. The Global North developed countries are at the forefront of ground-breaking technological innovations. Sub-Saharan African countries can draw numerous lessons from such practices, and comparative studies can help reveal such lessons and inform the tourism and hospitality industry of new technological trends. This volume further noted a dearth of literature on inclusive digitalisation. The inclusion of people with disability is critical to sustainable tourism development. Researchers can explore inclusive digitalisation in depth to inform industry practitioners on how they can achieve inclusivity in their operations.

13.6 Conclusion

This chapter provided a summary and concluding remarks for this volume. It is shown that sustainable tourism development is critical in this era of Sustainable Development Goals. There is no doubt that digitalisation contributes immensely to the sustainability discourse within the tourism and hospitality industry. The emerging findings and conclusions from each of the contributions in this volume are provided in this chapter. More importantly, lessons for key stakeholders on how to digitalise the tourism and hospitality industry for sustainable growth are also highlighted. Lastly, the chapter presents future research agendas drawn from contributions in this volume.

References

Abouzeid, N. (2022). The impact of digitalization on tourism sustainability: Comparative study between selected developed and developing countries. In *2nd international conference on advanced research in management, economics and accounting* (pp. 30–50). https://doi.org/10.33422/2nd.armeaconf.2022.02.10

Baker, J. (2012). The technology–organisation–environment framework. *Information Systems Theory, 1*, 231–245.

Dredge, D., Phi, G. T. L., Mahadevan, R., Meehan, E., & Popescu, E. (2019). *Digitalisation in tourism: In-depth analysis of challenges and opportunities*. https://ec.europa.eu/docsroom/documents/33163/attachments/1/translations/en/renditions/nativegeneral

El Archi, Y., Benbba, B., Zhu, K., El Andaloussi, Z., Pataki, L., & Dávid, L. D. (2023). Mapping the nexus between sustainability and digitalization in tourist destinations: A bibliometric analysis. *Sustainability (Switzerland), 15*(12). https://doi.org/10.3390/su15129717

Hoang-tien, N., Van Trai, D., & Quyet, N. X. (2024). The impact of digital transformation on tourism sustainable development: A case of SMEs in Vietnam. *International Journal of Entrepreneurship and Small Business*.

Leung, D., Lee, H. A., & Law, R. (2011). Adopting Web 2.0 technologies on chain and independent hotel websites: A case study of hotels in Hong Kong. In R. Law, M. Fuchs, & F. Ricci (Eds.), *Information and communication technologies in tourism 2011* (pp. 229–240). Springer-Wien.

Ndhlovu, E., & Dube, K. (2023a). Impact of COVID-19 on tourism in South Africa: Building resilience and prospects for recovery. In K. Dube, G. Nhamo, & M. P. Swart (Eds.), *COVID-19, tourist destinations and prospects for recovery volume one: A global perspective* (pp. 81–100). Springer Nature.

Ndhlovu, E., & Dube, K. (2023b). Challenges of radical technological transition in the restaurant industry within developing countries. *African Journal of Hospitality, Tourism and Leisure, 12*(1), 156–170.

Renn, O., Beier, G., & Schweizer, P. (2021). The opportunities and risks of digitalisation for sustainable development: A systemic perspective. *Gaia, 30*(1), 23–28.
Rodrigues, V., Eusébio, C., & Breda, Z. (2023). Enhancing sustainable development through tourism digitalisation: A systematic literature review. *Information Technology & Tourism, 25*(1), 13–45. https://doi.org/10.1007/s40558-022-00241-w
Shereni, N. C. (2023). Climate change disclosures by hotels in the global south. *African Journal of Hospitality, Tourism and Leisure, 12*, 1797–1805.
Shereni, N. C., & Chambwe, M. (2020). Hospitality big data analytics in developing countries. *Journal of Quality Assurance in Hospitality and Tourism, 21*(3). https://doi.org/10.1080/1528008X.2019.1672233
Shereni, N. C., Saarinen, J., & Rogerson, C. M. (2022). Website disclosures of sustainability practices in the hospitality sector: An analysis of hotel chains in Zimbabwe. *Tourism and Hospitality Management, 28*(3), 497–517.
Sørensen, F., & Baerenholdt, J. O. (2020). Tourist practices in the circular economy. *Annals of Tourism Research, 85*, 103027. https://doi.org/10.1016/j.annals.2020.103027
Tung, V. W. S., & Au, N. (2018). Exploring customer experiences with robotics in hospitality. *International Journal of Contemporary Hospitality Management, 30*(7), 2680–2697.
Woyo, E., & Ukpabi, D. (2022). Building resilient smart cities for sustainable urban tourism in Africa post-COVID-19 pandemic. In *Information and Communication Technologies in Tourism 2022* (pp. 256–270). https://doi.org/10.1007/978-3-030-94751-4_20
Zhou, Z. (2023). Human-wildlife tourism conflict: Towards an inclusive wildlife resources utilisation approach in southern Africa. *GeoJournal, 88*, 931–949. https://doi.org/10.1007/s10708-022-10667-8

Index

E. Ndhlovu et al. (eds.), *Tourism and Hospitality for Sustainable Development*, https://doi.org/10.1007/978-3-031-63069-9

The manufacturer's authorised representative in the EU is Springer Nature Customer Service Centre GmbH, Europaplatz 3, 69115 Heidelberg, Germany. If you have any concerns regarding our products, please contact ProductSafety@springernature.com

Printed and bound by CPI Group (UK) Ltd, Croydon, CR0 4YY
01/10/2026
02225965-0001